Emotion and Social Structu

T0227827

The past decades have seen significant advances in the sociological under-standing of human emotion. Sociology has shown how culture and society shape our emotions and how emotions contribute to micro- and macro-social processes. At the same time, the behavioral sciences have made progress in understanding emotion at the level of the individual mind and body.

Emotion and Social Structures embraces both perspectives to uncover the fundamental role of affect and emotion in the emergence and reproduction of social order. How do culture and social structure influence the cognitive and bodily basis of emotion? How do large-scale patterns of feeling emerge? And how do emotions promote the coordination of social action and interaction? Integrating theories and evidence from disciplines such as psychology, cognitive science, and neuroscience, Christian von Scheve argues for a sociological understanding of emotion as a bi-directional mediator between social action and social structure.

This book will be of interest to students and scholars of the sociology of emotion, microsociology, and cognitive sociology, as well as social psychology, cognitive science, and affective neuroscience.

Christian von Scheve is Assistant Professor of Sociology at the Department of Sociology and the Cluster of Excellence "Languages of Emotion," Freie Universität Berlin. His research focuses on the sociology of emotion, culture and cognition, social inequality, and social psychology.

Routledge Advances in Sociology

Emotion and Social Structures

The affective foundations of social order

Christian von Scheve

Routledge
Taylor & Francis Group

LONDON AND NEW YORK

First published 2013 by Routledge

2 Park Square, Milton Park, Abingdon, Oxon OX14 4RN
711 Third Avenue, New York, NY 10017, USA

Routledge is an imprint of the Taylor & Francis Group, an informa business

First issued in paperback 2017

British Library Cataloguing in Publication Data
A catalogue record for this book is available from the British Library

Library of Congress Cataloging in Publication Data
Scheve, Christian von.
   [Emotionen und soziale Strukturen. English]
   Emotion and social structures: the affective foundations of social order/
Christian von Scheve.
   pages cm. – (Routledge advances in sociology; 107)
   Includes bibliographical references and index.
   1. Emotions – Sociological aspects. 2. Social psychology. 3. Emotions and
cognition. I. Title.
   HM1033.S25613 2013
   302 – dc23   2012050654

ISBN: 978-0-415-67877-3 (hbk)
ISBN: 978-1-138-09428-4 (pbk)

Typeset in Times New Roman
by Florence Production Ltd, Stoodleigh, Devon, UK

Contents

Illustrations

Figures

Tables

Acknowledgments

The translation of this work was funded by Geisteswissenschaften International—Translation Funding for Humanities and Social Sciences from Germany, a joint initiative of the Fritz Thyssen Foundation, the German Federal Foreign Office, the collecting society VG WORT, and the Börsenverein des Deutschen Buchhandels (German Publishers & Booksellers Association).

This book is a thoroughly revised and abridged version of my dissertation research that appeared in German with Campus Verlag in 2009. I am grateful to Rolf von Lüde and Sighard Neckel, who supervised the work. I also thank Birgitt Röttger-Rössler and the fellows of the Research Group "Emotions as Bio-Cultural Processes" at the Center for Interdisciplinary Research (ZiF), Bielefeld University, where much of the dissertation was written.

Translated from German by Andrew Wilson.

Introduction

How can we reconcile the following and apparently contradictory observations: Individuals are embedded in societies whose structures and organizational patterns constitute the manifold frameworks that shape behavior, patterns of social interaction, and entire life courses. At the same time, individuals have a wide range of characteristics that make them—their beliefs, goals, desires—unique. This, in a nutshell, prompts the question of how society exists within the individual and how individuals constitute society. Sociology has investigated this question for decades by exploring forms and principles of socialization in relation to culture and social structure. The structure of societies though—of any social unit or aggregate—is not a fact ordained by nature but the product of interactions among a large number of actors. At the same time, however, societies demarcate the forms and conditions under which these actions and interactions take place. They provide opportunities and impose restrictions through the distribution of social and economic resources and the dissemination of norms and values.

These links between individual action and social structure have long been an established field of sociology, usually subsumed under concepts such as the "micro–macro link" or the "self–society dynamic" (Alexander & Giesen 1987; Howard 1991). Despite the many efforts and the progress that has been made, our knowledge on the connections between self and society is still not sufficient to provide satisfactory answers to fundamental questions concerning the individual processes and mechanisms underlying macro-social phenomena such as social change, social movements, or large-scale cooperation.

Remarkably, such questions are becoming increasingly topical and seldom has greater importance been attached to understanding these processes than in contemporary societies facing social change on global scales with unpredictable consequences for people's minds and behaviors. To make further progress in understanding these developments and consequences, the social sciences will have to develop advanced methods and concepts to illuminate the multifacetted connections between individual actors, the structures and qualities of social units, and the emergence and reproduction of social order.[1]

A fundamental question in this regard concerns the extent to which those determinants of social action that are traditionally considered to be most subjective and firmly rooted in individual minds and bodies are in fact systematically shaped by culture and society, and how, in turn, these determinants influence social action in a way that gives rise to large-scale social structural dynamics. Ever since Max Weber (1968), theories of social action have tended to emphasize intentional social action and its consequences for the emergence of social order, that is action based on more or less deliberate choices between alternative courses of action. Comparably little attention has been paid to those precursors of action that are outside conscious awareness, involuntary, and not reflexively accessible.

Weber had argued that *subjective meaning* is the key difference between human action and animal behavior. Consequently, this view—even if only implicitly—presumes that intention, volition, and motivation are the principal means of establishing links between the individual and society. Intentionality usually also denotes conscious reference to and knowledge of certain facts that constrain or enable social action, whether these are rational deliberations or normative obligations. This view, however, begs an intriguing question: What if there were also fundamental bi-directional links between non-conscious and pre-reflexive forms of action on the one hand and social order on the other? What if "the social" can also be located in the automaticity and pre-consciousness of everyday action, and this automaticity has a much profounder impact on the emergence and reproduction of social order than hitherto acknowledged?

As intriguing as this question might be, it is not a new one. The focus of much of sociological theory on the intentional qualities of action has been repeatedly called into question and has brought about a number of alternative approaches placing greater emphasis on the non-conscious and automatic aspects of social action, for instance practice theories (Reckwitz 2002; Schatzki 1996; Turner 1994). It would thus seem promising—not to say self-evident—to take a much closer look at disciplines which have been investigating precisely these foundations of action and behavior for quite some time. This seems all the more obvious because some of the disciplines in question have recently been gravitating towards the social sciences when it comes to explaining behavior in social contexts, chief among them certain strands of psychology, cognitive science, and also some branches of neuroscience. Most interestingly, these disciplines have marshaled theories and produced initial evidence that one factor in particular—namely *emotion*—could play a decisive role as a key intermediary between social action and social structure.

In going along with the common perception of emotions, it would seem that they exclusively belong to the subject areas of the behavioral sciences, since they are widely regarded as most deeply personal and subjective components of being. However, closer consideration reveals that emotions can—and indeed should—be analyzed from a sociological perspective; not simply to gain a more comprehensive understanding of emotions, but also to gain greater insight into their significance for sociality.

Some early sociological indications of this potential can be found in the works of Vilfredo Pareto (1935), Emile Durkheim (1915) or Georg Simmel (1901), to name but a few. Although these founding fathers of sociology did not produce a fully-fledged theory of emotion, they frequently allude to their significance for the individual and social phenomena. There are also most instructive references in the work of modern social theorists such as Norbert Elias (1994), Pierre Bourdieu (1984, 1992), Randall Collins (1975), and Anthony Giddens (1984) on the role of emotions in the emergence and reproduction of social order. However, as Barbalet (1998) argued, Weber's (1968) view of "affective action" as having only marginal significance for sociological analysis has remained paradigmatic for the sociology of the twentieth century. Weber held that sociology's primary field of enquiry should be the *ordered* life of actors in society and that "affective action," that is social action driven by passions and emotions, does not contribute much to explaining social order—and indeed mostly serves to undermine it.

In this book, I seek to advance the view that this prolonged marginalization of emotion is not only based on false premises but also puts modern sociology at risk of losing considerable explanatory potential, in particular when it comes to questions of structure and agency, of micro–macro linkage, and of self–society dynamics. Luckily, I am in good company. Although after Weber emotions remained a marginal category in sociology for a long time, the last four decades have seen an unprecedented rise in sociological research on emotion. Although the number of empirical findings still remains modest, this research has revealed the potential of emotions from various points of view, with regard to both a general theory of emotions and their significance for fundamental issues in sociology.

With a few exceptions (e.g., Barbalet 1998; Elster 1999; J. H. Turner 2007), however, many studies in the sociology of emotion have failed to take notice of the vast amount of research on emotions that has been accumulated in other disciplines over the past decades. As a result, much of the existing knowledge on human emotion has only played a minor role in sociological theorizing and research. One consequence is that general theoretical approaches to the sociology of emotion often have little connections with theories and findings of other disciplines, thereby forfeiting much of their potential explanatory power with regard to key sociological issues.

This book seeks to address these problems. Its central goal is to develop the argument that emotions play a key role in linking individual social action and the emergence and reproduction of social order. At first, this may seem paradoxical, but research in various disciplines engaged in emotion research strongly suggests that emotions account for a significant share of actors' daily, routine, and habitual actions and thus exert a decisive influence on the emergence and reproduction of social order and hence on the relationship between individual and society. Against this background, the following chapters will investigate emotions both conceptually and functionally as bi-directional mediators between social action and social structure. My primary goal is to elaborate their importance in terms of social theory, but I also aim at enriching our more general understanding of human emotion. Accordingly, emotion will be considered as both an "independent" and a "dependent" variable. On the one hand, I will investigate the ways in which social structures serve as a basis for explaining the elicitation and experience of emotions. On the other hand, I will look at how emotions influence social action and ensuing social structural dynamics.

Chapter 1 lays the groundwork by discussing micro-sociological perspectives on micro–macro linkages and the emergence and reproduction of social order, mostly those with an emphasis on knowledge and cognition. My aim in this chapter is to first develop a suitable framework for the subsequent analysis rooted in interdisciplinary emotion research and to ground emotions in established fields of sociological enquiries of micro–macro linkages. A brief review of sociological studies of emotion then highlights selected social structural theories, which are a cornerstone and frequent point of reference for the following analysis. Finally, I will develop a conceptual framework and working definition of emotion that informs the following investigation.

Chapter 2 then looks into the question of how and to what extent the generation of affect and emotion is structured by culture and society. First, I briefly outline some biological and cognitive foundations of emotion generation. In doing so, I highlight approaches that (a) emphasize the non-conscious and automatic aspects of emotion elicitation and (b) underline the fundamental social plasticity of these foundations. In sum, I hope to be able to sketch a picture of the *social structuration of emotion elicitation* that synthesizes biological and socio-cultural aspects.

Chapter 3 looks at the consequences of socially structured emotions for social action, with respect to decision-making in social contexts. In doing so, I first briefly review more established perspectives on the determinants of social action, especially those concerned with norms and rational choice. Subsequently, I elaborate on how emotions systematically influence these determinants and thus social action itself. I develop the notion of *affective action* as a form of social action that is markedly shaped by socially structured emotions and therefore decisively contributes to the reproduction of social order.

Chapter 4 then investigates the question of how (socially structured) emotions contribute to the development of robust patterns of social interaction. To this end, the first part of the chapter looks more closely at the expression of emotion and the recognition of such expressions by others. In the second part, social norms provide the basis for a twofold perspective on emotion-based social control. This form of social control is based on the regulation of emotion on the one hand, and on the enforcement of and compliance with various kinds of norms on the other hand. The final chapter briefly summarizes my main arguments, highlights unresolved questions, and suggests some avenues for future research.

1 Self, society, and emotion

To achieve an understanding of emotions as bi-directional mediators between action and structure, existing sociological paradigms seeking to explain these links on more traditional and established grounds are particularly helpful. The question of how individual action brings about and reproduces social structures at higher levels of aggregation, which at the same time constitute the opportunities and constraints for social action, has been characterized as the "Holy Grail" of sociology (DiMaggio 1991). During the past decades, North-American sociology in particular has developed theories that focus on micro-social processes such as face-to-face interaction, routinization, and categorization in explaining macro-social phenomena. These studies are characterized by a shift from normative–discursive and resource-based notions of social order towards a more cognitive concept thereof, taking a middle path between methodological individualism and collectivism that has been described as "methodological situationalism" (Knorr-Cetina 1981, p. 2).

As *cognitive sociologies* in the broadest sense, these approaches shall serve as a basis for developing a genuinely sociological perspective on emotion that can be linked to both existing emotion research in other disciplines and established accounts of (micro-)social order. This cognitive turn in sociology can be seen as an abandoning of Durkheim's conceptualization of normative moral obligations as *external* social institutions as well as of Parsons's internalization model, which indeed emphasizes internal over external control, but ultimately sees social action as realizing established normative ends. This turn has placed more importance on phenomena that are assumed to be *inherent* to individuals, such as language and cognition, which play a funda-mental role, for example, in the representation and interpretation of normative obligations. As far as action and behavior are concerned, it is primarily actors' *stocks of implicit or tacit knowledge* that are deemed most significant (Knorr-Cetina 1981, p. 4).

The crucial difference between this and more traditional normative perspectives on the formation of social order can be illustrated by looking at language. Just as the grammar of language is not consciously learned (nor is it hardwired, unless you are a Chomskyan), social and cultural patterns of behavior are—for the most part—not explicitly learned either. Rather, they are internalized and embodied during socialization. In contrast to, for instance, laws and legal norms, these behavioral patterns are not necessarily explicitly codified, and deviation is mostly subject to informal sanctions. In the case of the cognitive structures (or explicit and implicit knowledge structures) underlying social action, we are not dealing with the formal syntactic structures of linguistic competences, but rather with systems of meaning and significance that are *in principle* flexible and negotiable in social interaction. However, as later chapters will show, different forms of internalization and habituation may shape the cognitive structures and processes underlying meaning-making in a way that is similar to the rigid shaping of language. Affect and emotion play a crucial role in this shaping because they provide a distinct and bodily grounded level of meaning-making. Therefore, emphasizing cognitive structures and processes suggests a modified or even alternative stance to the emergence and reproduction of social order:

> Instead of a society integrated by common values and moral constraints, it is the *cognitive order of sense making and describing* which emerges from microscopic studies of social life. . . . In a sense, the problem of social order is redefined by turning the traditional approach to social order on its head. Social Order is not that which holds society together by somehow controlling individual wills, but that which comes about in the mundane but relentless transactions of these wills. The problem of social order has not only turned into a problem of cognitive order; it has also turned from a macro-level problem to a *micro-problem* of social action.
>
> (Knorr-Cetina 1981, p. 7)

From this perspective, the aim of this chapter can also be stated differently: Its purpose is to show that what we are dealing with is not simply a cognitive order of meaning-making, but also necessarily an affective and emotional order. To put it more forcefully, cognitive orders of meaning-making cannot exist or function in the ways proposed by existing accounts without corresponding affective orders. Ascribing a paramount role to cognitions and stocks of social knowledge in understanding social order suggests taking a closer look at those disciplines that have made cognition their very defining subject area: the cognitive sciences.

Interestingly, at least in the German-speaking academic community, virtually no convergence between sociology and the cognitive sciences can be observed to date (cf. Reichertz & Zaboura 2006), and also in the English-speaking countries such rapprochement is rare (cf. S. P. Turner 2002). This is all the more surprising since cognitive science usually considers itself a cross-disciplinary endeavor encompassing disciplines as diverse as philosophy, anthropology, sociology, neuroscience, and psychology. Many attempts to more extensively account for cognition in sociological analyses and to incorporate findings from cognitive science have come to grief, not least because of fears of a psychological or biological "reductionism" that (allegedly) cannot do justice to the complexities of the social world. This rather conservative view has been challenged emphatically by, among others, Paul DiMaggio (2002). He argues that psychology in particular can provide empirical tools that may help in making long-established sociological problems more readily accessible. Moreover, theories of social cognition can illuminate blind spots in sociological theory—DiMaggio cites the example of Bourdieu's (1992) concept of *habitus*. Last, they offer deeper insights into socially structured cognitive information processing.

In this vein, sociological social psychology has become an established area of research, both in psychology and (mainly North American) sociology. It is concerned primarily with the interaction between relatively stable cognitive structures (including, for example, values, attitudes, motives, and desires) and social structures (House 1981; Stolte *et al.* 2001). Sociological social psychology can be seen as a more psychological equivalent to micro-sociological theory. Unlike "standard" social psychology, however, sociological social psychology is not confined to investigating the influence of the social environment on individual minds. Rather it also seeks to contribute to the explanation of macro-social phenomena in the way more classical microsociology does. A basic assumption of sociological social psychology therefore is that stable social structures shape actors' cognitive structures, whose actions in turn reinforce these very structures and serve to reproduce corresponding social orders (House 1981; Stolte *et al.* 2001).

In this regard, however, Zerubavel (1997) emphasizes that such analyses are not a sufficient basis for satisfactorily explaining micro–macro linkages. He urges that the downplaying of individual differences in cognition must not be replaced by some sort of "cognitive universalism." The mere fact that cognition is highly flexible and shaped by the social environment does not mean that such shaping would lead to a general "homogenization" of cognition. Quite to the contrary, Zerubavel stresses that the cognitive differences and diversity that are found in different "thought communities," i.e. in distinct social units that shape cognitions in very specific ways, are of utmost importance to sociology (Zerubavel 1997, p. 9). In line with this view, I therefore advocate a *comparative approach* to cognition and cognitive structures that emphasizes the cognitive differences between as much as the similarities within distinct thought communities. This approach is, for example, mirrored by cognitive anthropology that has been concerned with intersections between culture and cognition for quite some time (D'Andrade 1981; DiMaggio 1997; Shore 1996; Sperber & Hirschfeld 2004). For cognitive sociology, however, the primary interest is not comparisons between cultures in an anthropological sense, but between cultures—as systems of cognitive structures and meaning-making—*within* societies (House 1981; Zerubavel 1997, p. 11).

Reflecting actors' embeddedness within these systems, cognitive structures determine how and what actors think and serve as a basis for this chapter's main argument in two ways: As facilitators of socially shared meanings and prerequisites for intersubjective understanding, cognitive structures necessarily require affective processes to be functional in the proposed way. In other words, the micro–macro-transcending effects ascribed to cognitive structures by certain branches of microsociology cannot be brought about without the involvement of affect and emotion. Second, these cognitive structures play a crucial role in the social structuring of emotion. If cognitions exhibit certain social and cultural patterns and are essential for the elicitation of emotions—as many emotion theories argue—then it necessarily follows that the resulting emotions exhibit similar patterns.

Aside from the question of how embeddedness into social structures and social orders affects thinking and feeling, a central problem for cognitive sociological theories is still that of scaling up micro-processes to macro-level structures, which, according to some, are the actual objects of sociology. In reviewing the possible mechanisms underlying the scaling of micro-social orders to macro-social structures, Knorr-Cetina (1981, p. 25 f.) locates existing accounts along two hypotheses: that of unintended consequences and that of aggregation.

The hypothesis of unintended consequences is not based on the assumption that macro-level phenomena can be completely reduced to micro-social events. Rather, it postulates that macro-level phenomena are a consequence of the unintended (and intended) consequences of micro-social episodes. These approaches argue in favor of the *emergence* of socially structured macro-level phenomena that are in principle irreducible to individual action. The aggregation hypothesis, on the other hand, maintains that macro-level phenomena are constituted by the aggregation and repetition of patterns of micro-social events. It is a rigorous interpretation of cognitive sociology and implies that macro-level phenomena can logically be derived from a corresponding analysis of all relevant micro-level elements.

In an attempt to resolve the dilemma produced by these hypotheses—which is that (a) all social action takes place in micro-social situations and that (b) these micro-level situations exhibit a tendency to become interlinked, producing unintended consequences that in turn affect social action—Knorr-Cetina (1981, pp. 30 ff.) suggests a third view: the "representation hypothesis." This hypothesis assumes that micro-level episodes' unintended consequences are first and foremost based on actors' perceptions and interpretations and thus primarily exist as *cognitive representations*. The dilemma is resolved in so far as social reality is indeed made up of micro-episodes, but at the same time macro-level phenomena can be taken into account as factors *endogenous* to these micro-level episodes (Knorr-Cetina 1981). Consequently, and regardless of their reification or objectification, social structures are in some way also *properties of actors* and are reproduced through structured ways of thinking, acting, and—as I argue—feeling.

This perspective is not confined to cognitive sociology and is indeed reflected in the view that social structures and the social orders of meaning-making are closely interlinked. Pierre Bourdieu's and Anthony Giddens's social theories, for example, are largely compatible with this perspective.

Bourdieu's explanation for the formation and reproduction of social structures is concerned primarily with internalized, habitualized, and socially differentiated patterns of perception, categorization, and interpretation that systematically shape (but do not completely determine) social action. Correspondingly, the *habitus* locates the structures of the social world within actors' structures of thought and perception (Bourdieu 1992; Lizardo 2004; Pickel 2005). To that extent, Bourdieu's notion of *habitus* can also be seen as the integration of a psychological and sociological structuralism characterizing it simultaneously as a "subjective" and "objective" structure. In a similar way, Giddens emphasizes the dialectic characterizing "subjective" and "objective" structures in his "duality of structure" account view (Giddens 1984). According to this account, social order arises from the reproduction of structures in social interaction and through the routinization of everyday action, which in turn has its origins in "practical consciousness." Strictly speaking, this practical consciousness does not in fact denote a state of consciousness, but expresses the possibility that everyday routine action frequently takes place without conscious awareness and is thus not discursively and reflexively accessible. Practical consciousness rather refers to the *general and socially distributed (tacit) knowledge structures* that are shared by large numbers of individuals within a society.

In summing up these perspectives on micro–macro linkages, I argue that there are several good reasons to—also—locate social structures *within* actors instead of merely viewing them as exogenous to individuals. According to cognitive sociology, these "internal" social structures are coupled to (or even constituted by) cognitive structures, which in its weakest form implies a systematic correspondence between social structures and cognitive structures that give rise to orders of meaning-making. Since cognitions, in turn, are fundamental to social action and behavior, it is plausible to assume that cognitive structures are causally implicated in bringing about regular patterns of social action, which in turn reproduce the "objective" and intersubjectively shared structures of the social world.

In conjunction with most of cognitive sociology and sociological social psychology, the term "cognition" here is primarily used to denote mental and symbolic contents and processes, i.e. representations such as beliefs, desires, and intentions as well as processes of perception, decision-making, and the storage and retrieval of information. This is not to say that I am taking a "dis-embodied" view on cognition. In contrast, I fully acknowledge the principles of embodied or even enactive cognition (Clark 2008; Noë 2004; Wilson 2002). However, my main interest in the embodied dimension of being lies in its affective components, and I am referring to the representational aspects of knowledge and cognition primarily to establish connections with existing accounts of micro–macro linkage in sociology. This is of course a purely analytical distinction, but it also allows me to contribute to an understanding of embodied cognition that incorporates affect- and emotion-specific bodily processes.

In this spirit, I will show that, to substantiate this view on the emergence and reproduction of social structures and social order, it is essential to account for emotions and affects, as some sociologists—most empathically Randall Collins (2004) and Jonathan Turner (2007)—have already indicated. As noted earlier, I will argue that emotions are a bi-directional mediator between social action and social structure and in many respects corroborate comparable cognitive links in terms of their *structural* and *structuring* potential. In view of the corresponding paradigms in sociological social psychology and cognitive sociology, I seek to shed light not only on what and how actors *think* as a function of social structure, but also on what and how they *feel*.

Micro–macro perspectives in the sociology of emotion

The micro–macro linkages outlined above have in part been addressed by social structural approaches to emotion and also by a number of symbolic interactionist or social constructionist accounts. Social structural approaches assign a critical role to emotions as intermediaries between action and structure and tend to refrain from exclusively treating emotions as either independent or dependent variables (Clay-Warner & Robinson 2008). Rather, they attempt to do justice to the manifold interdependencies characterizing the interplay of emotion and society. One of the best-known works in this respect is probably Theodore Kemper's (1978) *Social Interactional Theory of Emotions*. Kemper's starting point is that social structures can be characterized primarily in terms of two dimensions: status and power. In Kemper's model, the interpretation of specific configurations of status and power leads to characteristic physiological reactions, which in turn elicit specific emotions. Kemper's approach thus links elements of social structure, which are primarily manifested in social interaction, with the subjective interpretation of those elements, which give rise to certain physiological reactions that in turn produce specific emotions. Consequently, these emotions reflect the structural characteristics of social situations.

In a similar vein, Randall Collins (2004) explicitly addresses issues of micro–macro linkage. His theory is based essentially on the exchange of two resources, namely "emotional energy" and "cultural capital." The basic assumption is that actors are disposed to constantly strive to maintain or increase emotional energy, which can be understood as a form of gratification (Collins 2004). Consequently, actors tend to prefer those interactions that they expect to increase their emotional energy and to avoid those that are likely to produce losses. As a result, emotions become a resource and part of actors' preferences. The social structures and forms of stratification that are produced by these ritualized exchanges are facilitated by emotions in three ways: First, as an expressive modality, they are both distinctive elements and representations of individuals' experience and their embeddedness into social structures. Second, they motivate social action, mainly in view of changing allocations of emotional energy. The third function is brought about by the first two: The structure of a society that is stratified by a particular distribution of resources is in turn reinforced by the structuring effects of emotion.

Jonathan Turner's (2007) theory of emotion is in some respects similar to Collins's approach, but places greater emphasis on the evolutionary and neurophysiological foundations of human emotion. A basic tenet of Turner's approach is that accounting for the neurophysiological basis of emotion is crucial to better understand many sociological approaches to emotion: "Without biology, our explanations will be incomplete and seem rather shallow" (Turner 1999a, p. 101). Turner's reasoning relies on neuroscientific studies indicating that affect and emotion are to a great extent processed non-consciously, and that consciously experienced subjective *feelings*, as one component of an emotion, merely represent the tip of the iceberg (Turner 1999a).

The decisive question is to what extent the non-conscious and automatic processes involved in emotion elicitation, experience, and expression are of actual importance to sociology. According to Turner, emotions emerge during the course of social interaction when pre-existing general expectations are weighed against actual experience (Turner 1999b). These expectations are made up of demographic, structural, cultural, and transactional forces and condense in different concepts of self. These self-concepts are constituted in the course of social interaction, particularly through processes of role taking and role making, and determine actors' positions in the social space (Turner 2000). Face-to-face interactions are characterized by a number of needs which can be inferred from general expectations and include, for example, the need for group inclusion, ontological security, facticity, self-affirmation, and emotional and material gratification (Turner 1999b).

These emotionally relevant needs primarily contribute to the emergence and reproduction of social order: "people create, reproduce, or change social structures in terms of rewards or gratification" (Turner 1988, p. 357). Expectations, experiences, role taking, role making, and the satisfaction of needs all combine into specific patterns in the course of repeated social interactions which, through the structuring processes of regionalization, categorization, ritualization, stabilization of resources, normatization, and routinization shape macro-social reality (Turner 1988).

A further prominent structural account of emotion is developed by Jack Barbalet (1998), mainly with respect to class and social inequality. Barbalet argues that the structure of class relations determines the emotions of individual class members in that members constantly appraise these relations. When they are appraised as disadvantageous or as causes of deprivation, class members feel certain negative emotions, for example resentment (Barbalet 1992). Here, social structural configurations are mainly responsible for the unequal distribution of resources and asymmetries in power and reward allocated to members of different classes. These inequalities then become a major determinant of class-based resentment. Barbalet's approach is also remarkable because it explicitly aims to integrate structural and cognitive causes of emotion by assuming class-specific values, beliefs, and practices that system-atically impact class members' emotions. Moreover, Barbalet highlights that emotions are almost always precursors to social action, such that class-based emotions should lead to class-specific forms of behavior which in turn may affect the social structural conditions that originally gave rise to class-specific feelings (Barbalet 1992). He summarizes his position in stating that "emotions link structure and agency" and are a "necessary link between social structure and social actor" (Barbalet 2002, p. 3 f., italics omitted).

Besides these structural approaches, social constructionist and normative–discursive theories also provide valuable insights into the interplay between emotions and social order, here primarily understood as symbolic and norma-tive orders. Such approaches place particular emphasis on the social con-struction and normative orientation of emotions, on their valuation within society, as well as on practices that shape their experience and expression. The basic assumption is that the meaning, import, and functions of emotions are brought about by culture and society and are not hardwired in biological affect programs (Armon-Jones 1986; Averill 1980; Greco & Stenner 2007; Harding & Pribram 2009).

The constructionist paradigm ascribes crucial importance to social norms and rules. They constitute interpretive frameworks from which emotions emerge and provide actors with guidelines in adapting emotions and expres-sions of emotion to social expectations. It is no surprise, then, that social control is another important area of inquiry in the sociology of emotion (Hochschild 1983; Thoits 1990, 2004). From this perspective, the interplay of emotions and social order is primarily located within normative–symbolic orders and cultural practices, thereby valuably supplementing and extending the structural approaches.

The sociology of emotion gives insights into the ways in which social structure and emotion are linked and how emotions can possibly contribute to explaining the formation and reproduction of social order. Some of the theories in question also highlight that to advance our understanding of the complex interdependencies between social structure and emotion, we need to look at emotion research in other disciplines, primarily to gain insights into the possible mechanisms underlying this linkage. Some have argued that psy-chology and neuroscience have suggested avenues of connecting the basic principles of emotional functioning with actors' embeddedness into culture and society. At present, however, exploration of these possibilities is at the very beginning (Franks 2010; Hammond 2003, TenHouten 2007; Turner 2000).

Turner has criticized (1999a, 2007) the fact that the biological and psycho-logical bases of emotion are, for the most part, not adequately considered and conceptualized in sociology. His critique mainly concerns the failure to take into account the evolutionary origins of emotion and their significance for sociality in general and the importance of the physiological processes underlying emotion for social action and interaction. This is mirrored in the inadequacy of (even working) definitions of emotion often encountered in sociological studies. In some cases, definitions are simply non-existent while in others they are highly selective and sometimes even contrary to the available evidence (Turner 1999a, 2007). In countering these shortcomings, Turner in his own works gives detailed consideration of the evolutionary and biological basis of emotion (Turner 2000). However, he pays little attention to the cognitive structures implicated in emotion elicitation and to how they are linked to social structures and make up institutions. This criticism applies to many sociological theories of emotion, most of which are lacking fully developed models of the representation of social and cognitive structures.

At the other end of the continuum I locate normative–discursive approaches to emotion. Arlie Hochschild's (1983) groundbreaking studies on emotion work are an ideal starting point for investigating the role of emotion and emotion expression in face-to-face interaction, as I will prove in later chapters. Similarly, Eva Illouz's (1997) works vividly point to the power of discursive practices in shaping emotional experience and motivating emotion regulation in modern societies. However, these studies often imply an understanding of emotions that neglects their evolutionary origins and their bodily and physiological grounding, which in fact make up much of our emotional experience and are foundational in the (pre-reflexive) channeling of behavior. For the present purposes, I will therefore develop a working definition of emotion that aims at wedding its biological, cognitive, and social components.

Understandings of affect and emotion

What is an emotion? What is a feeling? And how are both elicited? Emotions are still at times defined only vaguely and implicitly on the basis of mundane understandings. This may be sufficient for everyday conversations over feelings and emotions, but poses limitations when we seek to identify the connecting lines between different scientific accounts and to integrate their various insights. To achieve this goal and to be able to bring together different views on the nature and culture of emotion, I will focus on those approaches that seem most promising in this respect and offer something close to an integrative interdisciplinary perspective. In doing so, I go along with Kappas (2002b) who in view of interdisciplinary approaches suggests that a unified and coherent definition of emotion based on assumptions about what is "right" and "wrong" is not (initially) necessary to make progress. He proposes to first find a minimal definition as a starting point of interdisciplinary research and to make a more comprehensive definition of emotion the goal of subsequent inquiries.

One definition that has gained considerable support among emotion researchers from different disciplinary backgrounds is reflected in "component" theories of emotion. These theories assume that an emotion is "an episode of interrelated, synchronized changes in the states of all or most of the five organismic subsystems in response to the evaluation of an external or internal stimulus event as relevant to major concerns of the organism" (Scherer 2005, p. 697).[1] This definition of an "actual" emotion that comes close to everyday understandings and experience is frequently referred to in the literature as an "emotion proper" (Goldie 2004, p. 94). As of yet, there is little agreement as to the exact number of constitutive components or the number of components that is sufficient to legitimately speak of an existing emotion proper. However, there are five core components that can be found in virtually every (componential) definition of emotion: physiological arousal, motor expression, subjective feeling, cognitive appraisal, and action tendency (Scherer 2005). According to this definition, emotions have an *episodic* nature. In other words, they are triggered by a specific event, are of rather short duration (seconds to minutes) and then fade away. The eliciting event must have a certain meaning and relevance for an individual and the emotion constitutes the reaction to this meaning. In this view, a discrete emotion (e.g., fear, anger, joy) is characterized by rather rapid and notable changes in the components involved and is further defined and specified by the nature and configuration of these changes.

Based on this working definition, various theoretical approaches to the nature and ontology of emotion can be identified that give different but not necessarily incompatible accounts of the links between emotions' biological basis and their socio-cultural shaping (see Engelen *et al.* 2008). Some of these approaches and debates center around the question of whether certain discrete emotions (i.e., specific configurations of changes in the involved components) are "innate" and universal across the human species and can therefore be inter-preted and intersubjectively understood shortly after birth, regardless of socialization and the social and cultural context. Proponents of this "basic emotion" approach hold that only a number of universal and "hardwired" features would allow for the many and rapidly occurring physiological and behavioral responses involved in certain emotions, which are assumed to be clearly distinct from other, so called self-conscious and social emotions (Ekman & Friesen 1975; Izard 1977; Johnson-Laird & Oatley 1992). Stronger accounts of basic emotions even suggest universality with respect to some of the events that elicit a basic emotion and some have argued that basic emotions are driven by distinct biological *affect programs* as their genetically developed cores (Griffiths 1997; Stein & Oatley 1992; Tomkins 1962). Ekman (1992a, p. 175) has listed a number of criteria that need to be met to qualify an emotion as "basic": a cross-culturally universal expression; presence in other primates; a distinct physiological event pattern; coherence between the autonomous nervous system activity and expressive behavior; quick onset of the emotion; short duration and an automatic appraisal process. According to these criteria, the basic emotions are anger, fear, sadness, happiness, disgust, and surprise. All other emotions are regarded as a mixture of various basic emotions (Ekman 1992a). From an evolutionary perspective, there is a fundamental difference in the function of the various categories of emotion. The role of the basic emotions is to maintain an organism's integrity and homeostasis, while social emotions ensure that individuals remain adaptable and flexible in a complex social environment (Cosmides & Tooby 2000; Turner 2000).

In criticizing the basic emotions approach, many have argued that there is not sufficient evidence to legitimately speak of discrete basic emotions or even discrete emotions as "natural kinds" (Barrett 2006). Russell (2003, 2009), for example, adopts a dimensional concept of affective experience and argues for an understanding of emotions not based on linguistic categories but rather along the experiential dimensions of valence (positive vs. negative) and arousal (high vs. low). This understanding is mirrored in the sociological Affect Control Theory of emotion (Heise 2010) that relies on the dimensions of the semantic differential (Osgood *et al.* 1957) and categorizes affective experience along the dimensions of valence, potency, and activity. Interestingly, these dimensions also seem to reflect a universal organizing principle for the mental representation of various components of emotion. Fontaine and colleagues (2007) have demonstrated that 144 common components of emotion, including physiological reaction, facial expression, and cognitive appraisal, map onto these dimensions (plus a fourth dimension representing surprise or unpredictability).

One of the most elaborated dimensional understandings of affect and emotion is proposed by Barrett and colleagues in her "psychological constructionist" account, sometimes also referred to as the Conceptual Act Theory of emotion (Barrett 2006, 2012; Barrett, Mesquita, *et al.* 2007; Barrett & Wager 2006). In a nutshell, Barrett argues that there is no conclusive evidence suggesting that "emotion" is in fact a "natural kind." Rather, she proposes that emotion concepts, which are represented in everyday life and scientific discourse by certain emotion words (for example, anger, fear, joy), are "nothing but" referrers to amalgamations of different processes that are also involved in various other behaviors and are not specific to emotion. Socially and culturally derived concepts and the verbal labels referring to patterns of physiological, psychological, and social processes in this perspective *count as* an emotion. Interestingly, Barrett (2012) also suggests that most of the physiological processes that are subsumed under certain emotion concepts are not specific to emotion.

Barrett argues that at the center of what are commonly referred to as emotions or emotional experiences are *core affects* that represent combinations of basic phenomenal (two dimensional) experiences of hedonic valence and arousal (see also Russell 2003, 2009). Variations in core affective feelings are linked to the specificity of emotional experience (Barrett & Bliss-Moreau 2009). In contrast to emotions, affect or affective feelings are assumed to be psychological primitives that "color" almost any sensory experience more or less vividly. Affective feelings are crucial elements of emotions, although they are by no means limited to emotions—think, for instance, of the affective feeling caused by pinning your finger with a needle. Affective feelings are also involved in moods, sentiments, attitudes, stereotypes, motivations, well-being, and many others (Barrett & Bliss-Moreau 2009). Core affect is "grounded in the somatovisceral, kinesthetic, proprioceptive, and neurochemical fluctuations" of the body, realized through specific neurophysiological states, and integrates sensory information from the external world with interoceptive information (p. 171 f.; see also Russell 2009). Importantly, affect also constitutes basic forms of meaning and meaning-making that are clearly different from prepositional and conceptual meanings. This has been shown, for example, by Osgood's work on the semantic differential and the affective meaning of words that are integral to Affect Control Theory (Heise 2010; Osgood *et al.* 1957). From this view, it is clear that discrete emotions are psychological and social constructions, and that basic or core affective reactions made-up of hedonic valence and arousal seem to be human universals. But still, even if basic affective reactions are ontologically universal to the human species, an interesting question lies in how far the conditions for their elicitation are shaped by culture and society.

In investigating the social structuration of emotion it therefore seems useful to not only look at discrete emotions as such (either "basic" or "complex"), as they are defined by componential approaches, but to also account for basic or core affective reactions that have equally important implications for behavior, as will be elaborated on in the following chapters. Moreover, this allows insights into the combinations of basic affective processes with higher order cognitions and socially and culturally shared conceptual (emotion) knowledge. This, in turn, promises to shed light on the characteristics of automatic and rapid, almost "instinct like" emotional reactions, without adopting reductionist positions (as, for example, in "affect program" accounts (see Tomkins 1962)) that cannot do justice to the complexity of the social world and its multi-faceted influences on emotion. I do assume that the mechanisms of core affectivity operate universally and are a necessary component of all emotions proper. However, even if a number of core affective reactions are biologically hardwired (fright and certain forms of fear, for example), I will emphasize that they are equally based on socially learned triggers and thus make an important contribution to the comprehensive social shaping of emotion. Therefore, in the following chapter I will discuss ways in which we can conceive of pre-reflexively elicited basic affective reactions not primarily as biological "givens," but as fundamentally shaped by the social embeddedness of actors.

2 Socially structured emotions

The view that emotions are a bi-directional link between social action and social structure is based on the premise that emotions are not just purely subjective and individual phenomena, as they are often portrayed. Rather, and in analogy to the relations between cognitive and social structures outlined in the previous chapter, I argue that the *generation* of emotion is *socially structured* as a consequence of individuals' embeddedness within stable social structures and social orders. This means that within social units characterized by specific systems of social order (for instance, groups, communities, or societies), emotional reactions towards certain classes of events—at least to some degree—tend to converge and to be coherent and in alignment with each other.

Therefore, social units are not only characterized by, for example, certain patterns of stratification, resource allocations, practices, and symbolic orders, but also by characteristic and relatively stable emotional "cultures" or "climates" (de Rivera 1992). In other words, certain emotions occur considerably more regularly in some social units than in others. This chapter, however, is not primarily concerned with the question of which particular emotions arise more often in which kinds of social units, but rather with the fundamental biological, cognitive, and social mechanisms that serve as a basis for this social structuring of emotion elicitation.

One of the earliest attempts at contributing to this question was made by Kemper (1978) in his *Social Interactional Theory of Emotions*. Kemper conceives of the elicitation of emotion as systematically dependent on social structural configurations and emphasizes the stability and durability of this dependency. For Kemper, a large class of human emotions arises from the subjective interpretation of one's position in the social hierarchy and the elements of social stratification. Given that patterns of stratification usually persist over long periods of time (decades to centuries) and that interpretations of one's position in the social hierarchy are based on values, beliefs, and desires, one would expect a corresponding stability in the emotions that arise in response to these structural configurations. Despite highlighting the subjective interpretation of relative social standing, Kemper also emphasizes the immediacy and automaticity of emotion that he locates in the tight coupling of social stimuli and emotion-specific physiological reactions (Kemper 1981, 1984).

It is precisely this coupling that has been at the center of extensive debate in emotion research and has been fuelled by differing views on the relationship between automatic, pre-reflexive physiological reactions and subjective interpretation in emotion elicitation. In one of the earliest accounts of this problem, James (1884) had advanced the argument that, contrary to wide-held belief, emotions are primarily determined by physiological reactions (such as blushing, sweating, or an increased heart rate).[1] According to this view, an individual experiences fear because of increases in heart rate and muscle tension—and not vice versa. Thus, for James, physiological reactions are the *cause* of an emotion and not the consequence. Although this account has been much debated, it has nevertheless become a starting point for many present-day emotion theories. It is particularly prominent in some branches of neuroscience and psychology (for example basic emotion theories and affect program models) that see the core of emotions in hardwired biological or physiological processes and leave little (conceptual) room for the influence of culture and society on basic mechanisms of emotion elicitation.

James's view was probably most vividly challenged by the cognitive turn in the social and behavioral sciences. Many considered physiological reactions as too coarse and undifferentiated to account for the broad variety in human emotional experience. Rather, following the now classical experiments by Schachter and Singer (1962), cognitive interpretations were thought to be the connecting link between a stimulus, the ensuing physiological arousal, and the subsequent emotion. Today, a similar position is adopted by advocates of *appraisal theories* of emotion that focus not primarily on hardwired physiological reactions, but instead on the subjective interpretation of an event.

More recently, a number of integrative theories of emotion elicitation have pursued a middle way between these antagonistic positions and attempted to compensate for the respective shortcomings. In view of basic emotion theories and physiological models, these shortcomings are often found in explanations of complex social emotions. Conversely, some of the cognitive theories fall short in explaining the automatic and involuntary elicitation of basic affective responses. Nevertheless, both are relevant in uncovering the social structuration of emotion, although at first glance, the explanatory potential of the cognitive approaches might seem more substantial. Consequently, this chapter argues that the social shaping of emotion can take place equally well at the level of these basic affective responses and their physiological underpinnings. To achieve a better understanding of how this shaping is achieved, a closer look at the bodily mechanisms involved in eliciting basic affective responses can be instructive. This includes investigating how distinctions can be made between events that elicit affective responses and those that do not at the level of "early" automatic and non-conscious affective processing.

The answer to this question might pave the way to understanding the social shaping and structuring of emotions at levels previously thought to be biologically determined. If core affective reactions were mostly hardwired in the human brain, then "higher" cognitive processing could only contribute to their differentiation and categorization as well as their coupling with symbolic and conceptual knowledge in the generation of emotions proper. However, if it can be shown that basic affective reactions are systematically shaped by culture and society, this fosters an understanding of the social structuring of emotion as not only taking place in the realm of conceptual knowledge and symbolic interpretations, but also on the level of early and pre-reflexive affective reactions.

This chapter will therefore first outline a view on the bodily foundations of emotion that promotes an understanding of the fundamental social plasticity of emotion elicitation, primarily at the level of human neurophysiology. Subsequently, the chapter elaborates a perspective on the cognitive components of emotion that allows the linkage of sociological conceptions of the social nature of knowledge and cognition (as outlined in the previous chapter) with the elicitation of emotions.

Neurophysiological foundations of emotion elicitation

To shed light on the neural correlates of emotion processing, I will primarily draw on studies and reviews from *affective* and *social neuroscience* (Davidson 2003c; Lieberman 2007) as well as on models and theories that have been widely received and discussed in the field of the *neuroscience of emotion* (see Adolphs & Damasio 2000; Gainotti 2000). The seeking of those elements of the bodily bases of emotion that render emotion elicitation susceptible to social and cultural influences is primarily guided by hints at such a social plasticity within eminent theories and empirical studies. Many neuroscience models account for social influences on emotion in terms of evolutionary adaptation, although ontogenetic development and the effects of socialization are increasingly recognized. A large proportion of existing studies focuses on higher cognitive components of emotion as the primary locus of social influences. Although studies concentrating on the neural correlates of basic affective responses do acknowledge the influence of culture and society, these acknowledgments remain mostly implicit. Consequently, the remainder of this chapter categorizes emotion elicitation into rapid, automatic, and pre-reflexive processes on the one hand, and controlled and reflexive processes on the other hand.

William James (1884) is frequently cited as the beginning of modern psychological research on emotion. The answer to his question "What is an emotion?" is still a matter of debate, because it attributes a key role to physiological reactions in the process of emotion elicitation:

> Our natural way of thinking about ... emotions is that the mental perception of some fact excites the mental affection called emotion, and that this latter state of mind gives rise to the bodily expression. My thesis

on the contrary is that *the bodily changes follow directly the perception of the exciting fact, and that our feeling of the same changes as they occur is the emotion.* Common sense says, we lose our fortune, are sorry and weep; we meet a bear, are frightened and run . . . The hypothesis here to be defended says that this order of sequence is incorrect, that the one mental state is not immediately induced by the other . . . , and that the more rational statement is that we feel sorry because we cry, angry because we strike, afraid because we tremble, and not that we cry, strike, or tremble, because we are sorry, angry, or fearful, as the case might be.

(James 1884, p. 189 f.)

James's hypothesis marks one of the first explicit attempts at discussing the links between physiological and mental processes in the elicitation of emotion. Contrary to the then prevailing view that emotions were the consequence of a more or less deliberative evaluation of an event, James postulated that it was primarily physiological reactions that determined the subjective phenomenal perception of an emotion. According to this view, the differentiation of discrete emotions was directly dependent on discernible patterns of physiological arousal. Consequently, James was placing physiological rather than mental activity at the center of the emotion process—a position that has since been vigorously debated and has certainly had its critics, particularly in social constructionism. However, the model has also been criticized in other quarters, primarily with regard to the potential of physiological arousal to differentiate the various emotions. For example, Cannon (1927), although placing physiology at the heart of emotion, too, asserted that physiological arousal is not sufficiently differentiated and, above all, emerges too late in the process of emotion generation to explain the wide variety of phenomenal experiences related to emotion (see Cacioppo *et al.* 2000, p. 175).

To better understand the behavioral black box between stimulus and emotional reaction and to provide some basic explanations for the mechanisms involved, early neuroscientific investigations used lesion studies to examine the functions of certain brain areas in human and animal behavior. By looking at changes in overt behavior and cognitive capacities, researchers made inferences on the functions of the parts of the brain that were damaged or lesioned. Results showed that even animals with extensive or complete lesions of the cerebral cortex still showed emotional behavior with characteristic reactions of the autonomous nervous and motor systems. However, the more the cortex receded, the more clearly changes in behavior began to appear: Animals were much more easily provoked and the emotions they displayed were usually inappropriate to the situation, extraordinarily intense, and increasingly undirected or undifferentiated (LeDoux 1996, pp. 79 ff.).

These studies led to the development of a dual-circuit model of emotion elicitation that to a large extent remains in use today: If, in addition to certain cortical areas, deeper and phylogenetically older brain structures were damaged or lesioned, these deficiencies often lead to the complete loss of affective reactions. These mostly subcortical brain areas—first described by James Papez (1937) in the so-called "Papez Circuit" and later termed the "visceral brain" (or the "limbic system") by Paul MacLean (1952)—broadly consist of structures clustered around the brain stem and forming the inner border of the cortex. They include the hippocampus, which plays an essential role in memory, the hypothalamus, which controls basic biological functions and is responsible for the body's hormonal state, and the amygdala, which plays a key role in evaluating the affective salience of stimuli (LeDoux 1996, pp. 92 ff.; Roth 2003, p. 256 f.; see Figure 2.1).[2]

Early lesion studies had identified two systems whose functions seemed to be of particular importance for the experience of emotion. On the one hand, subcortical circuits that are responsible for the emergence of basic affective reactions, and, on the other hand, paralimbic and cortical areas implicated in differentiating, regulating, and attributing these reactions. Following on from these studies, it is now widely believed that complex and highly differentiated human emotionality is fuelled by a relatively rudimentary "affect system" that also exists in other mammalian species (Barrett & Bliss-Moreau 2009; Duncan & Barrett 2007). On the one hand, this system is made up of limbic subcortical structures that produce affective reactions and hardwired responses to certain features of stimuli, having evolutionarily important functions related to homeostasis, bodily integrity, and behavioral control. Processing in these circuits is assumed to be rapid, automatic, and occurring largely outside conscious awareness.[3]

In addition to these limbic components, certain paralimbic and cortical areas of the brain play an important role in the further differentiation, categorization,

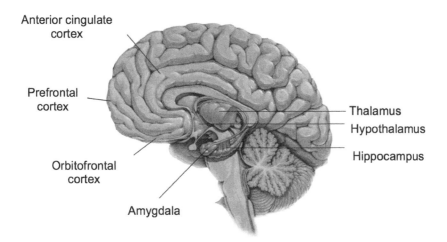

Figure 2.1 Key cortical and subcortical brain areas implicated in affective processing.

Source: Adapted from Lewis (2005, p. 179).

and control of affective reactions. These neural circuits, which include areas of the frontal lobes (the prefrontal and orbitofrontal cortex) and the anterior cingulate cortex, also play an important role in the cognitive representation of affect, the influence of affect on executive functions, for example in action planning and implementation, and the integration of affective and cognitive information, which also includes conceptual emotion knowledge. The more social and self-conscious emotions requiring complex representational capabilities, such as shame, pride, and envy, are inconceivable without the involvement of these areas.[4] There is a broad consensus in the literature that an evolutionary perspective is essential in understanding emotion, so that the results of studies on non-human animals can reasonably be taken into consideration. This applies in particular to the functions of the subcortical affect system, but also in part to the paralimbic and cortical parts of the system.[5] Also, neuroimaging studies have provided evidence that emotion and cognition are subserved by specialized and highly interactive neural circuits, although opinions differ as to the implications of these interactions and the exclusiveness of processing.[6]

Social science accounts of emotion have frequently dismissed the importance of the limbic affect system for their areas of inquiry because it allegedly leaves no room for ontogenetic development and the adaptation to culture and society and therefore "opens the floodgates" to "biological determinism." These reservations can be countered by noting that one of the outstanding features of emotion elicitation in humans is the capacity for learning, relearning, and adaptation—not only in view of complex social emotions, but also regarding basic affective responses.

In the following sections, I will outline the structure and operation of these two systems in more detail. In doing so, a key concern will be at what point, in what brain areas, and through what mechanisms the process of emotion elicitation may be susceptible to social and cultural shaping.

The basic affect system

The affect system, much like other perceptual systems, is primarily designated to transform stimuli into a meaningful "motivational metric" and to initiate appropriate behavioral responses (Cacioppo & Gardner 1999; Cacioppo *et al.* 2004). The structures implicated in this system "represent crucial components of a network that bind sensory stimulation from inside the body to that coming from outside the body, and in so doing each gives the other informational value" (Barrett & Bliss-Moreau 2009, p. 173). One of the most prominent models of the affect system is based on fear conditioning studies conducted by Joseph LeDoux (1996). LeDoux was primarily interested in the question of which brain areas ultimately imbue a stimulus with affective meaning and significance. He focused on the processing of auditory information in non-human animals and subsequently traced the information processing pathways of conditioned (i.e., learned) fear responses to auditory stimuli.

LeDoux showed that conditioned fear responses remain intact even after certain cortical areas of the brain had been removed. Animals in his studies failed to show conditioned fear responses, however, after suffering lesions to subcortical structures, in particular the auditory thalamus or certain areas of the midbrain. LeDoux (1996, pp. 150 ff.) drew two key conclusions from his observations: First, the thalamus seemed to play a central role in the processing of emotionally relevant sensory stimuli, at least as a relay station to other regions. Second, sensory cortices are, in principle, dispensable in fear conditioning or, more precisely, in coupling sensory perception and affective reaction. Further investigations revealed that the thalamus transmits information to other subcortical regions, in particular the amygdala, which, upon damage or removal, completely impaired fear conditioning.

LeDoux concluded that the amygdala is obviously a central neuronal structure implicated in attributing affective meaning to sensory stimuli. The amygdala consists of several nuclei that receive information not only from the thalamus but also from the sensory cortices. Furthermore, it passes information on to several other subcortical areas, for example the hypothalamus, the brain stem, and the midbrain, whose principal function is to initiate physiological and motor reactions (such as stiffening, change in blood pressure, and the release of hormones) (LeDoux 1996, pp. 155 ff.). Other studies have highlighted the fact that that amygdala activity is not exclusively related to fear responses. Rather, its function is to direct attention to sources of sensory stimulation, to determine "whether exteroceptive sensory information is motivationally salient" (Lindquist *et al.*, 2012, p. 130), and to initiate physiological reactions preparing the organism for appropriate behavioral responses (Anderson & Phelps 2001; Barrett, Bliss-Moreau, *et al.* 2007; Cacioppo *et al.* 2000; Davidson 2003c; Murray 2007; Phan *et al.* 2002; Sander *et al.* 2003).

The function of the auditory cortex in fear conditioning, on the other hand, is primarily to differentiate and identify the conditioned stimulus. It is true that lesions to the auditory cortex do not prevent fear conditioning but they do seem to remove the ability to discriminate between comparable stimuli —only one of which, for example, may be combined with an aversive, fear-inducing stimulus. In the case of lesions to the auditory cortex, both stimuli would trigger fear reactions in much the same way, because the amygdala receives no input from the sensory cortices and relies on relatively fuzzy information from the thalamus. According to this view:

> the Beatles and Rolling Stones . . . will sound the same to the amygdala by way of the thalamic projections but quite different by way of the cortical projections. So when two similar stimuli are used . . ., the thalamus will send the amygdala essentially the same information, regardless of which stimulus it is processing, but when the cortex processes the different stimuli it will send the amygdala different signals.
>
> (LeDoux 1996, p. 162 f.)

In view of these studies, LeDoux's (1996) overall conclusion is critical to the main argument of this chapter. He suggests that emotional learning in general and fear conditioning in particular do not necessarily require the participation of cortical areas, which in the case of the sensory cortices are responsible for precisely identifying and differentiating stimuli and in other cases for higher cognitive abilities such as conceptualization and semantic categorization. This means that the basic subcortical affect system performs certain functions, including associative learning, which in earlier models were believed to be processed in cortical areas only and that imply flexible adaptation to the social environment.

Based on these findings, LeDoux developed a "dual-path" model of affective processing (see Figure 2.2). At the neural level, emotional learning takes place along two pathways: On the one hand, the short and quick "low road" is characterized by the subcortical processing of information and a more or less direct link between the sensory thalami and the amygdala. On the other hand, the "high road" provides for significantly more detailed and differentiated but considerably slower processing of information by recruiting cortical networks (LeDoux 1996, pp. 163 ff.). Subcortical processing relies on simple and fuzzy representations of a stimulus, and its principal characteristic is the rapidity of processing that comes at the cost of error-proneness (quick and dirty). Basic affective reactions to certain stimuli are produced in these subcortical areas *before* a stimulus can even be consciously perceived. This function may explain, for example, why we often experience certain feelings without being consciously aware of what has triggered these feelings in the first place—as in the proverbial "gut reaction." One of the noteworthy features of this

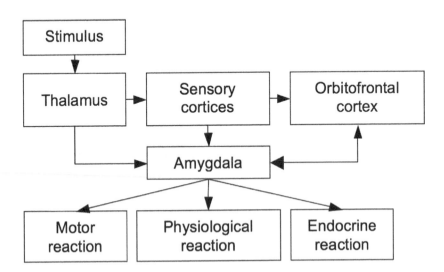

Figure 2.2 Simplified "dual-path" model of the basic affect system.
Source: Adapted from LeDoux (1995, p. 215) and Rolls (2002, p. 4447).

architecture is that it allows for regulatory control mechanisms located in cortical areas. When detailed representations of a stimulus are transmitted to the amygdala via the "high road," then—under certain circumstances—initial behavioral responses initiated by the amygdala on the basis of thalamic projections may be overridden. In these cases, the role of the cortical areas is correcting inadequate affective reactions rather than initiating them (LeDoux 1996).

Edmund Rolls (1999, 2002) adopts a similar but in some respects more detailed approach to explaining the elicitation of affect and emotion. He defines emotions as states elicited by instrumentally rewarding or punishing stimuli. Rewarding stimuli are those for which an organism is willing to invest energy. Punishing stimuli, conversely, are those on whose avoidance an organism is willing to spend energy. Punishing and rewarding stimuli act as "reinforcers" since they influence the probability that certain behaviors will occur. "Instrumental reinforcers are stimuli that, if their occurrence, termination, or omission is made contingent upon the making of a response, alter the probability of the future emission of that response" (Rolls 1999, p. 61). The actual reinforcers are the somatosensory consequences of a stimulus, such as taste, smell, tactile information, and pain (Rolls 2004, p. 18). Accordingly, emotions can be differentiated by means of various stimulus–reinforcer associations.

From a social science perspective, Rolls's distinction between primary and secondary reinforcers is of particular interest. Primary reinforcers are *innately* rewarding or punishing stimuli such as taste, touch, smell, and certain visual stimuli. Secondary reinforcers are acquired through learning and associated with primary reinforcers. They may represent a wide range of different kinds of information, such as certain visual patterns, symbols, or abstract social concepts, such as money (Knutson & Bossaerts 2007; O'Doherty *et al.* 2001). At first glance, the concept of secondary reinforcers might resemble the mechanisms of classical conditioning. However, its function is fundamentally different from that of stimulus–response learning. In such learning, conditioned stimuli elicited specific behavioral reactions. Secondary reinforcers, however, elicit affective reactions that allow for flexible and adaptive behavioral responses (Rolls 1999, p. 62; 2002, p. 4444).

In the process of emotion elicitation, first the reward value of a primary reinforcer is determined. Associations are then made between hitherto neutral stimuli and primary reinforcers. According to Rolls, this associative learning takes place primarily in two brain structures, the amygdala and the orbitofrontal cortex (OFC). In order for primary and secondary reinforcers to be processed, stimuli are initially identified—in contrast to the dual-path model—at the level of categorical objects in the sensory cortices. Subsequently, their reward value is determined in the amygdala and the OFC, in which Rolls locates representations of primary reinforcers and the corresponding learning processes for secondary reinforcers (which themselves are in turn represented in the sensory cortices).

These parts of the brain appear to be especially important in emotion and motivation not only because they are the parts of the brain where the primary (unlearned) reinforcing value of stimuli is represented in primates, but also because they are the regions that learn pattern associations between potential secondary reinforcers and primary reinforcers. They are therefore the parts of the brain involved in learning the emotional and motivational value of stimuli.

(Rolls 2000, p. 185)

The amygdala's task here seems to be the integration of information from the primary and secondary reinforcers (stimulus–reinforcement association learning), and its primary function is to determine and represent the reward value of a stimulus. According to this view, the amygdala attributes aversive or appetitive significance to sensory stimuli by associating them with primary or secondary reinforcers (Rolls 1999, p. 102). In contrast to LeDoux's model, Rolls's theory relies more heavily on pre-processed information from the sensory cortices, but does not fundamentally challenge the notion that the amygdala reacts to information from other subcortical areas, such as the thalamus (Rolls 2000, p. 185).

This is no surprise, given that LeDoux focused on animal studies and reactions to rudimentary sensory information. In humans, complex configurations and patterns of stimuli, such as voices, faces, symbols, and other sensory information that require more extensive processing in the sensory cortices to be behaviorally meaningful, are much more important (Rolls 1999, p. 104 f.). Because of the potential diversity of secondary reinforcers, which include complex social and cultural information, Rolls's model provides valuable insights into how the elicitation of emotion can be socially shaped in a systematic fashion at the level of basic affective responses.

In addition to generating core affective feelings, the basic affect system is also implicated in producing responses that are essential for the action tendencies accompanying emotion. In particular, these include influences on cognitive processing and executive functions in action planning and decision-making, which will be investigated in greater detail in the following chapter. These behavioral and motivational tendencies are thought to be mediated by two neural systems which have been described as *approach* (or attack) and *avoidance* (or withdrawal) systems (Davidson & Irwin 1999) and broadly correspond to Rolls's (1999) criteria of positive and negative reinforcers that also exhibit motivational and behavioral characteristics.

These two systems are supposed to translate a wide range of stimuli into a common "motivational metric" along a positive and a negative evaluative dimension (Cacioppo & Gardner 1999, p. 199). Importantly, in this "evaluative space model" appetitive or aversive affective behavior is initiated only in combination with a third dimension representing learned "affective dispositions" towards certain classes of stimuli based on combinations of positive and negative affectivity (Cacioppo & Berntson 1999, p. 135; Cacioppo *et al.* 2004,

p. 228). In contrast to the models outlined previously, Cacioppo and colleagues assume that these evaluative dimensions are each based on specific neural systems comprising, for example, nucleus accumbens for positive, amygdala for negative, and hypothalamus for both kinds of evaluations (Cacioppo *et al.* 2004, p. 230).

These views on the elicitation of basic affective reactions suggest that there is significant overlap in a number of prominent neuroscientific emotion models that at the same time emphasize the rapid and automatic generation of affective responses in subcortical brain networks and their susceptibility to social and cultural shaping (Davidson 2003a; Ochsner & Barrett 2001). It should be noted that, although the affective reactions produced within milliseconds by this affective system are relatively undifferentiated and, presumably, merely bipolar within a positive–negative continuum, they still have far-reaching consequences for more complex emotional reactions, cognitive processing, and social action. What has also become clear is that affective reactions are not further differentiated let alone semantically categorized until they are further processed in cortical networks.

An interesting question then is what sorts of representations the basic affect system can fall back on? According to the theories outlined above, areas related to the "low road" must be able to form representations of the (social) environment that allow, however approximately, for the identification and categorization of a stimulus. To return to LeDoux's example: The Beatles and the Rolling Stones may sound very similar to the amygdala, but the Rolling Stones and Beethoven do not. One possibility may be that core affective reactions are not produced by representations of particular categorical objects, such as a wild and dangerous animal or a villain, but rather that certain properties of stimuli, such as size, movement, or appearance, can be categorized and represented already in subcortical areas of the affect system (Damasio 1994, pp. 131 ff.). In what follows, I will address the questions of what information can possibly be represented and differentiated in these areas and which representations have to be processed in cortical areas and thus cannot be processed with the same rapidity and automaticity (LeDoux 2000, pp. 171 ff.).

Complex and social emotions

In addition to the output of the basic affect system, emotions proper, that is fully-blown emotions comprising components such as action tendencies, cognitive appraisals, and facial expressions, require the involvement of further cortical brain areas to more precisely categorize eliciting stimuli and to associate them with conceptual knowledge and (autobiographical) memories in the process of meaning-making. To uncover the possibility of the social structuration of these emotions proper, accounts of the neural basis and processing of more complex and "social" emotions can be instructive.

Emotions in everyday situations arise not only in response to the perception of rudimentary sensory information, but more often than not constitute reactions to behaviors, signs and symbols, thoughts, ideas, plans, and memories (Solomon 2004, p. 16 f.). In these cases, operation of the basic affect system has to be supplemented by areas of the brain capable of processing more complex cognitive information, for example, the sensory cortices, certain areas of the associative cortex, the anterior cingulate cortex and prefrontal or orbitofrontal areas (Barrett, Mesquita, *et al.* 2007; Damasio 1994; Ochsner & Barrett 2001; Roth 2003; Teasdale *et al.* 1999).

The experience of complex (social) emotions requires establishing connections between thoughts, ideas, and memories (or conceptual and learned representations, more generally) with basic affective reactions that have become associated with these representations. These kinds of emotions "occur once we begin experiencing feelings and forming systematic connections between categories of objects and situations, on the one hand, and primary emotions, on the other" (Damasio 1994, p. 134; italics omitted). Conscious perceptions and higher cognitive activity that are processed in sensory and associative networks in turn initiate further processing in prefrontal, in particular orbitofrontal, areas. Importantly, these parts of the human brain are supposed to store what Damasio (1994, p. 102) has termed "dispositional representations." These are acquired representations that associate certain classes of stimuli with certain patterns of affective and physiological responding. As such, dispositional representations store the patterns of neural activity of other brain regions, including those of the basic affect system. Areas of the prefrontal cortex in which these representations are located transmit information to the amygdala and other subcortical areas, which then initiate affect-specific physiological and motor reactions (Damasio 1994, pp. 136 ff.; Damasio *et al.* 2000). As such, dispositional representations act as "somatic markers": They recruit activity of the basic affect system that is characteristic of responses not only to rudimentary stimuli (as with "primary reinforcers" or "primary emotions"), but also to other kinds of conceptual, higher-order stimuli, for example thoughts, memories, and signs and symbols. Crucially, dispositional representations therefore elicit what can best be described as "gut reactions," that is, a more or less constant stream of feelings that becomes attached to a multitude of social situations. From a sociological point of view, dispositional representations and somatic markers can be seen as the affective components of social practices.

The role of the frontal lobes in these processes is well documented. It has been shown that impairment of the OFC gives rise to notable changes in social and emotional behavior and deficiencies in decision-making (Bechara 2004; Kringelbach & Rolls 2004). In stimulus–reinforcement association learning, the OFC also plays a fundamental role, since it represents the rewarding and punishing values of primary and secondary reinforcers. This region is particularly well suited to this function, since it receives and integrates information from the sensory cortices, which represent reinforcers' characteristics (Kringelbach & Rolls 2004, pp. 347 ff.; Rolls 2000, p. 183 f.; Rolls 2004, p. 18; Wood & Grafman 2003). Interestingly, the OFC seems to be highly flexible with regard to the learning and relearning of stimulus–reinforcement contingencies.

> Decoding the reinforcement value of stimuli, which involves for previously neutral (e.g., visual) stimuli learning their association with a primary reinforcer, often rapidly, and which may involve not only rapid learning but also rapid relearning and alteration of responses when reinforcement contingencies change, is then a function proposed for the OFC
>
> (Rolls 2004, p. 18)

This function is particularly important in complex social situations, in which rewarding and punishing values are constantly changing and stimulus–reinforcer contingencies regularly have to be learned anew. This is the case, for example, in social exchange requiring cooperation and coordination. In contrast, the subcortical affect system, and in particular the amygdala, seems to function much less efficiently in response to alterations in stimulus–reinforcement contingencies and the relearning or reversal of stimulus–reinforcement associations (Rolls 1999, 2004).

A good example of the rigidity of such associations are phobias, which are regarded as virtually irreversible and can at best be overlaid by new associations (LeDoux 1996, p. 225–266). In reacting to altered reward values, the OFC is not dependent on the amygdala as an interface for other subcortical structures, since it can directly initiate behavioral responses via the nucleus caudatus, ventral striatum, and the hypothalamus (Kringelbach & Rolls 2004; Rolls 1999). Distinguishing between primary and secondary reinforcers is also important for the elicitation of complex social emotions. Because the OFC has advanced representational capacities and is thought to represent propositional, conceptual, and non-(somato)sensory reward values, such as money, cultural values, and other resources, these more abstract representations can act as secondary reinforcers in associative learning (O'Doherty *et al.* 2001; Schaefer & Rotte 2007). Against this background, certain characteristics of social structures and social order, such as status and power allocations, norms, and values most probably have a reward potential for associative learning and thus act as secondary reinforcers.

Moreover, the strong reciprocal connections between orbitofrontal areas and parts of the subcortical affect system, in particular the amygdala, grant the OFC cognitive control precedence over the affect system: The strongly developed connections suggest that one of the functions of the prefrontal cortex is to modulate and inhibit amygdala activity (Davidson 2004; Lieberman *et al.* 2007; Rolls 1999). Further support for such a control function comes from studies on emotion regulation showing inverse activation patterns for the amygdala and the prefrontal cortex in the intentional regulation of emotion (Ochsner & Gross 2005).

However, before intentional emotion regulation can occur, affective reactions have to be subjectively experienced and categorized, not only at the level of phenomenal feelings but also at the conceptual level. Subjective affective feelings are not only a necessary component of an emotion proper, but also implicated in cognitive information processing and behavioral control. The subjective experience of an affective state requires working memory that retains phenomenal feelings in consciousness and is accessible to manipulation and other volitional cognitive activities. Studies on decision-making suggest that this "affective working memory" is located in the (ventromedial) prefrontal cortex and is based on representations of elementary positive or negative affective states in the absence of the actual elicitors (Davidson & Irwin 1999). Moreover, impairments in these regions have been shown to prevent the integration of affective and cognitive information in decision-making (Damasio 1994, 2003).

Besides the prefrontal cortex, parts of the cingulate cortex also play an important part in the elicitation and regulation of emotion. In view of the emotion regulatory functions of the OFC, however, it is unclear how exactly the need to control or inhibit an affective response is determined. Some have suggested that such a need arises when cortical and subcortical circuits provide contradictory representations of sensory input. In such cases, the anterior cingulate cortex (ACC) is assumed to be implicated in behaviors involving the monitoring and evaluation of one's internal state and behaviors regarding events that prompt the need for deliberate changes in behavior (for example in view of errors, uncertainty, or conflicting goals and desires) (Ochsner & Barrett 2001).

By virtue of its diverse connections with other cortical and subcortical brain areas, the ACC—much like the somatosensory cortices—represents the body's physiological state and so contributes to controlling not only cognitive processes and motor behavior, but also the internal milieu. The ACC together with the OFC therefore plays a critical role in evaluating social situations characterized by, for instance, contingency, uncertainty, specific affective expectations, or underspecified reward values (Roth 2003, p. 280). This in particular applies to situations in which other actors behave aberrantly or unpredictably, such as when violating social norms. Taken together, the ACC contributes to emotion elicitation and regulation in cognitively demanding situations and by interfacing higher cognitive functions with basic affective reactions. Among other things, its relevance for complex social emotions is closely tied to the attribution of affective states to an event and the accounting for conceptual—and most likely socially and culturally shared—emotion knowledge in this process (Ochsner & Barrett 2001; Ochsner & Gross 2005; Phan *et al.* 2002; Teasdale *et al.* 1999).

Generally, the preceding paragraphs have shown that certain regions of the frontal lobe, in particular the orbitofrontal and ventromedial prefrontal cortex, play a central role in the elicitation of complex emotions by linking basic affective reactions to higher-order cognitions and conceptual knowledge. These areas of the brain supplement and interact with the subcortical affect system in order to facilitate the elicitation of emotions proper. In this view, the basic affect system produces positive or negative phenomenal feelings, based on sensory input or input from cortical brain areas, as well as other physiological components of emotion. Cortical areas are mainly involved in triggering (or modulating) the affect system based on dispositional representations and conceptual knowledge and by providing socially and culturally derived categorizations of affective reactions. Also, affective feelings are retained in affective working memory in cortical structures and exert considerable influence on action and decision-making. Cortical areas also play an important role in the modulation and inhibition of basic affective responses.

The view of emotion elicitation that emerges from these perspectives strongly relies on the representation of complex (social) information. This is evident from the representational capabilities of secondary reinforcers as well as from dispositional representations and their role in emotion generation. The consequences for the social structuration of emotion elicitation are evident when accounting for the cultural origins and the social sharing of (conceptual) knowledge and its role in associative learning and conditioning, which take place in both cortical and limbic systems. The following section examines the ways in which this potential for the social structuration of emotion is further realized at the level of cognitive representations and information processing.

Cognitive foundations of emotion elicitation

In answering his question "What is an emotion?," James (1884) suggested that the subjective perception of a physiological reaction is the actual emotion. James argued that perception, physiological reaction, and emotion were more or less directly linked. Some of the neuroscientific research outlined previously is supportive of this position, although James's critics raised concerns not only regarding the links between physiology and emotion but also with respect to his very definition of emotion. Schachter and Singer (1962) from their studies concluded that cognitions must be critically involved in emotion elicitation and constitute the bridge between rather unspecific physiological arousal and the experience of discrete emotions. Thus, in addition to specific patterns of physiological arousal, cognitions become necessary components of an emotion proper. One particularly necessary (but not sufficient) component is supposed to be a dedicated cognitive process that causally links physiological arousal with the evaluation of an event (LeDoux 1996, p. 47 f.; Reisenzein 1983, p. 240). Although Schachter and Singer's (1962) study could hardly be replicated, it is still regarded as the catalyst for most contemporary "cognitive" theories of emotion, in particular a variant known as "appraisal theory," and also for some sociological approaches to emotion, first and foremost Kemper's (1978) status and power theory.

Generally, within emotion research, the relation between cognition and emotion is one of the most vividly debated issues. One of the main disputes regards the question of whether cognitions are necessary for emotion elicitation or even a constitutive component of emotion. The now classic dispute between Lazarus (1984) and Zajonc (1980) is indicative of the widespread disagreement on the relation between (higher) cognitive and automatic, non-conscious processes in emotion elicitation:

> Over the past few decades, scholars of emotion theory ... have put forward the hypothesis that affective processing does not depend on controlled cognitive processing. That is, they proposed that organisms are able to determine whether a stimulus is good or bad without engaging in intentional, goal-directed, conscious, or capacity demanding processing of the (evaluative attributes of the) stimulus. Rather, affective processing could occur automatically. Such automatic affective processing was believed to have an important impact on subsequent cognitive processing and behavior
>
> (De Houwer & Hermans 2001, p. 113).

Contrary to some radically cognitivist positions that have their origins in Schachter and Singer's experiments and conceive of emotions as "nothing more" than particular configurations of cognitive states (e.g., Solomon 1976), Zajonc (1980) had argued and empirically demonstrated that emotional evaluations of stimuli and corresponding affective preferences (e.g., liking vs. disliking) are formed even when the relevant stimuli are not consciously perceived and deliberately processed. He concluded that "preferences need no inferences," i.e. that affective reactions towards stimuli do not rely on cognitions. Lazarus (1984), on the other hand, argued that Zajonc's position was based on an excessively narrow understanding of cognition. He held that cognitions, at least in the sense of basic information processing, were essential for emotion elicitation. In this view, information processing does not have to take place consciously and need not be limited to "higher" cognitive processes, such as planning or rational deliberation. For Lazarus, therefore, the primary trigger of emotions was the cognitive appraisal of an event, which, in the form of automatic and non-conscious appraisals, could also explain Zajonc's findings.

Looking at several neuroscience models of emotion—some of them illustrated above—it seems clear that basic affective reactions can indeed occur as the result of rapid and automatic appraisals of an event without a stimulus being consciously perceived and processed:

> some, perhaps many, of the things we do, including the appraisal of the emotional significance of events in our lives and the expression of emotional behaviors in response to those appraisals, do not depend on consciousness, or even on processes that we necessarily have conscious access to.
>
> (LeDoux 1996, p. 65)

Many of the existing studies do indeed support Zajonc's claims but at the same time nevertheless contradict the view that cognition plays a decisive role in emotion elicitation (Parrott & Schulkin 1993, p. 45).

Early on, Leventhal and Scherer (1987) had seen the cognition vs. emotion debate primarily as a "semantic controversy" about *definitions* of what cognitions and emotions are. Their insights and arguments gave rise to a number of integrative approaches that sought to better bring together the psychological and physiological components of emotion and to emphasize the interaction between various levels of information processing (Barnard & Teasdale 1991; Clore & Ortony 2000; Ellsworth 1994; Lewis 2005; Robinson 1998; Scherer 1993a; Smith & Kirby 2000). Today, most neuroscientific research rests on the assumption that affect and cognition are not processed by entirely separate neural systems, but rather by strongly interacting and non-exclusive networks.

> We now understand that emotion is comprised of many different sub-components and is best understood not as a single monolithic process but rather as a set of differentiated subcomponents that are instantiated in a distributed network of cortical and subcortical circuits.
>
> (Davidson 2003a, p. 129)

Integrative approaches emerging from this debate usually distinguish between controlled and explicit as well as automatic and implicit processes of emotion elicitation (a distinction already tentatively made in the previous section) and define cognition relatively broadly, as including not only higher cognitive but also more basic processes such as perception, pattern recognition, and information storage and retrieval (Frijda 1994; Robinson 1998).

In a similar fashion, this view has been marshaled by LeDoux (1993, p. 62):

> If we define cognition narrowly (as the highest levels of thought and intellect) then cognition is not necessary for emotion. On this there is little disagreement, for even the most ardent cognitive theorists of emotion accept that emotional reactions can be elicited without the involvement of the highest levels of thought and intellect . . . If cognition is defined broadly as information processing, then emotion must be dependent upon cognition. Sensory processing, even by peripheral receptors, is information processing and therefore emotion must be dependent upon information processing and thus upon cognition.

This perspective is well suited to combine findings related to the cortical and subcortical processing of basic affects and emotions with cognitive models, since it brings together automatic and non-conscious processes in emotion generation, in particular the elicitation of basic affective responses with more complex cognitions. In this regard, Leventhal and Scherer's (1987) component process model (CPM) of emotion that distinguishes between "sensorimotor," "schematic," and "conceptual" processing has played a pioneering role.

Cognitive process models like the CPM can be seen as a further development and specification of early appraisal theories of emotion, which traditionally focused on the cognitive structure of evaluations and subjective interpretations in the process of emotion generation (for discussions, see Barnard & Teasdale 1991; Clore & Ketelaar 1997). However, already the pioneers of appraisal theory had suggested that appraisals were not necessarily conscious and deliberate, so that in principle they are specialized processes also involved in bringing about very basic affective reactions and, to some extent, may be a general *modus operandi* of the affect system (Arnold 1960, p. 172).

The neuroscience research reviewed at the beginning of this chapter relates to a rather narrowly defined set of emotion components that are indeed necessary but in no way sufficient for the elicitation of an emotion proper. Some appraisal theories allow incorporating more complex cognitive structures and processes underlying large parts of everyday human emotion (which cannot be reduced to affective reactions alone) into this system. Accounting for more complex cognitions is crucial since "an emotion is not what happens in the first 120 milliseconds of arousal . . . An emotion is not the initial neurological reaction" (Solomon 2004, p. 19). As I have shown, the basic affect system is undoubtedly critical to the sociological study of emotion. However, it only sheds light on a fraction of phenomena commonly referred to as emotions. As argued in the previous section, the affect system is essential to any explanation of core affective reactions, such as the fright experienced by an unexpected noise or the pleasure gained from sweets. But this system is not sufficient to understand emotions proper and their full-blown socio-cultural precursors and repercussions, such as the shame one feels when caught red-handed committing theft or the sorrow experienced when seeing rising unemployment rates and a declining economy.

The overwhelming majority of emotions experienced in everyday life do not arise in situations of bodily emergency or following the acute perception of certain stimuli innately associated with deeply rooted affective responses. Rather "a very large class of emotions results from real, imagined, or anticipated outcomes in social relationships" (Kemper 1978, p. 43). And only seldom does an emotion—in contrast to some basic affective reaction—result from a hypothetical "baseline" of arousal, but usually from an ongoing stream of affective experiences and complex cognitions such as representations, evaluations, imaginations, interpretations, and plans. Before an actual emotion develops, individuals do not find themselves in some sort of "affectless state": Simply the anticipation of one's first date triggers joyful expectations, while the memory of a long-ago accident might fill one with unease, and the prospect of a job interview might give rise to feelings of anxiety. Nico Frijda has very aptly summarized this view:

> emotions result from meanings, and meanings, to a large extent, from inferred consequences or causes . . . A majority of emotional stimuli derive their emotional impact along these lines: those of generalization

or, rather, abstract thought, of application of rules and general knowledge schemas, of generalized and normative expectations. That impact . . . has little to do with having experienced aversive or pleasurable consequences accompanying a particular kind of stimulus.

(Frijda 1986, p. 310)

Keith Oatley (1992, pp. 19 ff.) has argued along similar lines, stating that a particular stimulus is not in itself sufficient to reliably trigger a distinct emotion in a given individual. On the other hand, a specific stimulus, such as an unexpected noise for example, is sufficient to reliably trigger basic affective responses, such as shock or surprise, in virtually anybody. Emotions proper, Oatley suggests, are rather based on evaluations of an event against the background of cognitive representations such as goals, beliefs, and previous experiences:

I may suddenly feel frightened if the vehicle in which I am travelling seems to be heading for an accident. I evaluate a perception in relation to my concerns for safety, though not necessarily consciously. This may be the common experience in such situations. But a person confident that an accident would not occur, perhaps the one who is driving, or the one who is unconcerned about personal safety at that moment, may not feel fear.

(Oatley 1992, p. 19)

These views are mirrored in the inclusion of "higher" cognitions in many definitions of emotion, for example:

* the intentional directedness of an emotion towards an object (one is annoyed *about* an event, afraid *of* an examination or embarrassed *by* a derogatory remark) (Goldie 2000, p. 16–27);
* the motivational aspects of the action tendency accompanying an emotion (one tries to eliminate the reasons for an annoyance or to restore one's standing after an embarrassing situation) (Frijda 2004; Oatley 1992, p. 24); or
* the assessment of potential resources to cope with a situation that triggered an emotion and with the emotion itself (Scherer 1984, 1999).

Accordingly, most "cognitive" theories emphasize the importance of goals, beliefs, desires, and other mental states in generating emotions, many of which are of cultural origin and are socially shared among many actors (see Frijda *et al.* 2000; Ortony *et al.* 1988).

The following section discusses the precise ways in which cognitive structures, which are presumably processed and represented mainly in cortical brain areas, contribute to the elicitation of complex emotions. I will also seek an understanding of how complex emotions proper can be elicited automatically and non-consciously, much in the same way as basic affective reactions. This analysis shall primarily help to uncover links between well-established sociological views on the social structuration of knowledge and cognition on the one hand, and emotion elicitation on the other hand. Subsequently, therefore, I will discuss to what extent cognitive processes are in fact *socially constituted* and or *socially shared* and how they might interact with other levels of information processing relevant to emotion elicitation, in particular those in subcortical brain areas. I will also seek to clarify at what stages of emotion elicitation subcortical and cortical processes can access which kinds of representational formats at what degrees of complexity (e.g., bodily, iconic, propositional) to better understand the interplay of automatic, non-conscious, and deliberative, conscious processing in emotion. These interactions do not only promise insights into the extent to which social structures and social order may shape various components of emotion, but also provide potential indications in view of the role of emotions in the structuration of social action and interaction and, ultimately, the *reproduction* of social order.

Social cognitive structures of emotion elicitation

Looking at the manifold interactions between cognition and emotion, one of the most interesting questions relates to the role cognition plays in the generation of emotion.

> Whether an event elicits an emotion in an individual and, if so, what emotion (joy, sorrow, fear etc.) and with what intensity depends on how that individual interprets the event—in particular, how he or she appraises it relative to their goals and desires.
>
> (Reisenzein 2000, p. 117; author's translation)

At first, this perspective on emotion elicitation could be taken from a sociological textbook. However, it reflects a basic principle of one of the most prominent psychological theories of emotion, namely appraisal theory. The many parallels with sociological and social constructionist approaches are best seen regarding the notion of the interpretation of relevance. To date, however, there have been virtually no attempts in the sociology of emotion to account for the multitude of empirical findings and concepts inspired by appraisal theory. What might be the reason for this lack of interest? As noted earlier, the majority of sociological theories either start from the assumption that emotions are social constructions and are produced by subjective interpretations and processes of meaning-making, or they understand social construction as a process that modulates or regulates emotions once they have arisen. The first variant is not fundamentally different from the assumptions underlying appraisal theory. It is all the more surprising, therefore, that some more recent sociological theories often distance themselves from the interpretative and constructionist paradigm and rely instead primarily on the findings of affective neuroscience (for discussions, see Franks 2010; Franks & Smith 1999; S. P. Turner 2007).

One reason for this distancing is to be found in—justified—criticism of "over-socialized" conceptions of emotion:

> if emotions depend on the interpretation of the situation, it seems that all who define the situation similarly ought to experience the same emotion. The problem, in part, comes down to whether or not it is possible to have a standard set of categories for defining situations which will link them logically and empirically with emotions. . . . The social constructionists provide no overarching framework of situations to which one may refer for the prediction of emotions.
>
> (Kemper 1981, p. 352 f.)

It is precisely this task of providing a comprehensive framework for interpreting situations and linking interpretations to emotions that appraisal theories aim to accomplish: "All situations to which the same appraisal pattern is assigned will evoke the same emotion. . . . [A]ppraisals start the emotion process, initiating the physiological, expressive, behavioral, and other changes that comprise the resultant emotional state" (Roseman & Smith 2001, p. 6 f.; italics omitted). The idea of appraisal as the primary trigger of emotions goes back to Magda Arnold (1960) and Richard Lazarus (1968) and denotes the evaluation or assessment of an event in light of an individual's goals, beliefs, desires, and intentions as wanted or unwanted, pleasant or unpleasant, familiar or unfamiliar. The term "events" denotes both internal processes, such as memories or representations, and external phenomena, such as objects, actors, situations, or actions (Ortony *et al.* 1988, p. 18 f.). According to appraisal theory, appraisals establish a relationship between the internal criteria for evaluating an event on the one hand and the properties and characteristics of an event on the other hand:

> appraisals are inherently *relational* . . . Rather than exclusively reflecting either the properties of the stimulus situation or the person's dispositional qualities, appraisal represents an evaluation of the stimulus situation as it relates to the person's individualized needs, goals, beliefs, and values.
>
> (Smith & Kirby 2001, p. 124)

Besides the appraisal of events on the basis of their (intrinsic) positive or negative valence (primary appraisals), most theories add additional criteria that contribute to the further differentiation of the quality and intensity of an emotion (secondary appraisals) (Lazarus 1991a). Most of these do not directly relate to the event but rather to certain consequences or properties derived from it, such as subjective coping potential, the attribution of agency, or the probability of its occurrence. Another assumption common to most appraisal theories is that different emotions are accompanied by different patterns of appraisal; in other words, each distinct emotion is triggered by a corresponding distinct appraisal pattern. This means that appraisals in fact *precede* emotions and are causally implicated in triggering them—they are not epiphenomena that somehow accompany an emotion.[7]

By and large, appraisal theories face two challenges (see Reisenzein 2000): They have to make certain assumptions about the precise nature of the appraisal process as well as about the structure and contents of appraisal processes. The former primarily concern the temporal sequences and characteristics of the various modes of information processing involved in appraisal, and the latter concern the specificity of the various dimensions and patterns of appraisal and the cognitive structures (or contents) that constitute one end of the relational character of appraisal.

Most theories and empirical studies have focused on the structural properties of appraisal. Their primary objective is to analyze the "cognitive structure of emotions" (Ortony *et al.* 1988), uncovering which cognitions are involved in emotion elicitation in what way (Frijda 1986; Roseman 2001; Scherer 1984; Smith & Ellsworth 1985). To that end, most theories highlight a certain number of appraisal dimensions that relate to specific properties of appraised events. These include, for example, intrinsic valence, goal conduciveness, motive congruence, probability of occurrence, coping potential and resources, and attribution of responsibility (e.g., Kemper 1978; Parkinson 1997; Roseman *et al.* 1996; Scherer 1984).

Furthermore, appraisal theories need to specify which appraisal patterns, that is which combinations of the various dimensions of appraisal, give rise to which distinct emotions. In this regard, Reisenzein (2000) distinguishes between the explanatory power of such relations with regard, first, to the *internal* structure of an emotion (i.e., the composition of relevant appraisals) and, second, to the *external* structure of emotion, which reflects the relations between different emotions. In what follows, I will be focusing primarily on the internal structure.

As far as the conceptualization of the contents of appraisal is concerned, a number of different approaches can be identified (see Parkinson & Manstead 1992; Roseman & Smith 2001). Some authors take the view that appraisals are based on *discrete* categorical elements; in other words, events are evaluated by reference to the constitutive elements of an actor's cognitive structure or motivational concerns, i.e. on the basis of beliefs, goals, and desires.[8] Although these structures are dynamic to varying degrees, it is often assumed that appraisals resulting from these cognitions produce *categorical* results. For example, a goal is either achieved or not, an event meets or does not meet expectations, an action conforms to or deviates from prevailing norms. Thus the results of an appraisal can be clearly differentiated from each other, as can the resultant emotions (Kemper 1978; Oatley 1992; Ortony *et al.* 1988; Roseman 1991, 2001; Smith & Lazarus 1993). An alternative (though strongly related) way is taken by theories that conceptualize appraisals in *dimensional* terms. According to this view, emotion-eliciting events are evaluated on the basis of a number of appraisal dimensions, such as attention, pleasantness, certainty, goal congruence, agency, and legitimacy (Smith & Ellsworth 1985). Some of the most prominent appraisal dimensions and categories are summarized in Table 2.1.

To better understand the ways in which appraisals are supposed to operate and to uncover the potential for the social shaping of both appraisal contents and processes, I will discuss two appraisal theories in more detail: First, the categorical model developed by Ortony *et al.* (1988), which is referred to in many disciplinary contexts and empirical studies and, second, Scherer's (1984, 1999) dimensional approach, which has gained some prominence in psychological emotion research. Both theories can be seen as supporting the argument of the social structuration of emotion for two reasons: First, Ortony and colleagues' (1988) approach makes explicit reference to cognitive structures and thereby unveils the extent to which appraisals are shaped by cultural and societal factors. In addition, further developments of the theory (e.g., Clore & Ortony 2000) supplement the predominantly structural approach of the original account with a detailed model of the cognitive processes involved in emotion generation. Second, Scherer's dimensional appraisal model is widely recognized and has received considerable empirical support. Importantly, several empirical studies have highlighted the cultural variability of certain appraisal structures as well as the universality of a number of appraisal mechanisms (e.g., Scherer 1997). Furthermore, this approach is firmly rooted in an elaborated account of the information processing involved in emotion elicitation (Leventhal & Scherer 1987; Scherer 1984).

Ortony and colleagues (1988) have outlined the various cognitive structures underlying appraisal patterns of a number of discrete emotions. They conceive of emotions as "valenced reactions to events, agents, or objects, with their particular nature being determined by the way in which the eliciting situation is construed" (Ortony *et al.* 1988, p. 13). Events are relevant by virtue of their consequences and desirability, agents by virtue of the praiseworthiness of their actions, and objects by virtue of their appealingness. The ways in which a particular situation is appraised depends on an agent's current goals, standards, and attitudes. The desirability of an event is determined by the hierarchy of an individual's goals; the praiseworthiness of an action is judged by the

Table 2.1 Appraisal criteria suggested by different appraisal theories

Frijda (1986)	*Ortony & Clore (1988)*	*Roseman (1991)*	*Smith & Ellsworth (1985)*
Change	Unexpectedness		
Valence	Appealingness		Pleasantness
Open/closed	Desirability	Motive consistency	Goal/path obstacle
Controllability	Agency	Agency	Agency
Value relevance	Blameworthiness		Legitimacy

Source: Adapted from Scherer (1993b, p. 327).

prevailing standards and norms; and the appealingness of an object is evaluated by reference to attitudes or predispositions (Ortony *et al.* 1988, p. 58).

As far as goals are concerned, Ortony and associates distinguish "active-pursuit goals" ("things one wants to get done") from "interest goals" ("things one wants to see happen") and "replenishment goals," a sub-class of active pursuit goals which are cyclical in nature (e.g., paying one's rent) (Ortony *et al.* 1988, p. 41; italics omitted). "Standards" refer to social or moral norms as principles and guidelines for social action. In the appraisal process, norms reflect the behavioral expectations of other actors and are particularly relevant in the elicitation of complex emotions, such as shame or guilt. Attitudes are defined as predispositions to like or dislike certain objects (as in certain tastes or aesthetic preferences).

Goals, standards, and attitudes thus form the cognitive-structural "backbone" against which objects, events, and actions are appraised and corresponding emotions emerge. Emotions in this model are based on three basic types of evaluations of the consequences of events (pleased/displeased), the actions of agents (approving/disapproving), and the properties of objects (liking/disliking) (Ortony *et al.* 1988, p. 33). These basic evaluations give rise to three categories of emotions: event-based, agent-based, and object-based emotions. Event-based emotions are distinguished from one another with respect to the focus of the event (who is affected), the likelihood of its occurrence, and the degree of conformity to expectations. Agent-based emotions relate to the degree of approval of an action for which the agent is held responsible or for which responsibility is attributed to another individual. Object-based emotions are elicited by objects that have been assessed as attractive or unattractive.

In contrast to this account of the "cognitive structure of emotion," Scherer's (1984, 1999) model focuses on specific appraisal *dimensions*. He suggests that emotions are triggered by a fixed sequence of successive appraisals, the "stimulus evaluation checks" (SECs) (Scherer 1984, 1999). These appraisals take place in a specific order, with each level of the appraisal process being based in part on the results of the previous level (Scherer 1999). SECs usually begin with appraising the novelty of a stimulus. This rudimentary appraisal is based, for example, on instinctive reactions such as the orienting or startle reflex and can presumably be fully processed, in neural terms, on subcortical "low roads." However, it is also concerned with less sudden impressions and may encompass, for example, comparison of perceptual input with conceptual representations. The second SEC is related to the affective valence that is *inherent* to a stimulus and contributes to its phenomenal experience (intrinsic pleasantness). This SEC is similar to the appraisal of an object's "attractiveness" in Ortony and colleagues' model (1988). Despite the emphasis on valence that is inherent to a stimulus, this appraisal can be based on acquired as well as innate representations (Leventhal & Scherer 1987, p. 15; Scherer 1984). The third SEC evaluates stimuli in light of their congruence with an individual's concerns (goal/need conduciveness). Scherer distinguishes between goal conduciveness and intrinsic valence, because even events that are intrinsically pleasant can, under certain circumstances, have negative effects on goal attainment. Conversely, an appointment with the dentist may imply intrinsically unpleasant affective reactions but generally serve the goal of maintaining one's health (Clore *et al.* 1994, p. 352). Goal conduciveness consists of four secondary appraisals (Leventhal & Scherer 1987, p. 15): relevance, conformity to expectations, goal conduciveness, and urgency. The fourth SEC determines an agent's coping potential and includes four secondary appraisals: cause of and responsibility for an event, controllability of consequences, available resources, and ability to adapt to uncontrollable consequences. The fifth and final SEC evaluates an event's norm- and self-compatibility and is further specified into two categories. On the one hand, it establishes whether an individual's own actions or those of others conform to social norms and expectations. On the other hand, it establishes the extent to which an action is compatible with personal standards that are part of self and identity. Table 2.2 illustrates which combinations of appraisals are presumed to lead to which discrete emotions.

Both of these appraisal models show that distinguishing between dimensional and discrete appraisals is at best a question of perspective. In both theories, appraisals establish relationships between an event (an actor or object) and agents' stocks of implicit and explicit knowledge, beliefs, desires, goals, and norms—most of what is constitutive of individuals' mental and physical lives, reflecting their historicity, sociality, and culturality (see Smith & Kirby 2001, p. 124 f.). Consequently, appraisal dimensions can also be seen from an actor-centered perspective and from the underlying cognitive structures: The novelty of a stimulus, for example, is determined by comparing existing representations with actual sensory input. Also, one's coping potential

Table 2.2 Predictions of emotions based on SECs (simplified)

	Joy	Fear	Anger	Sadness	Disgust	Shame	Guilt
Novelty/expectancy							
Expectedness	open	low	open	open	open	open	open
Intrinsic pleasantness							
Unpleasantness	low	open	open	open	high	open	open
Goal/need conduciveness							
Goal hindrance	low⁻	high	high	high	open	open	low
Coping potential							
External causation	open	external	external	open	external	internal	internal
Coping ability	medium	low⁻	high	low	open	open	open
Compability with standards							
Immorality	open	open	high	open	open	open	high
Self-consistency	open	open	low	open	open	low⁻	low⁻

Source: Adapted from Scherer (1999, p. 773).

Note: High = very high; low– = very low; open = unspecific.

is based on resources (economic, cultural, social, psychological, and physio-logical) which are either available or not. Appraising the norm congruence of an event presupposes that norm and behavioral expectations are in some way (cognitively) represented. In similar ways, many existing appraisal theories and the criteria of appraisal they postulate require different sorts of (cognitive) representational structures against which perceptual input is compared in appraising an event (e.g., Frijda 1986; Roseman 1991, 2001; Smith & Ellsworth 1985) (see also Table 2.2).

This understanding of dimensional appraisal accounts (in addition to the existing categorical accounts) in terms of mental representations and cognitive structures underlying appraisal is crucial for the argument of this chapter. If appraisals can be conceptualized not only on a number of evaluative dimen-sions, but also on the basis of representations and cognitive structures, this allows for directly relating the societal and cultural origins of knowledge and cognition to the social structuration of emotion elicitation. Importantly, there is no need to reconceptualize the actual appraisal processes and mechanisms —which are assumed to be largely universal—but rather to demonstrate the systematic social and cultural shaping of cognitive structures that most appraisal models treat as a given. Interestingly, Ortony and colleagues (1988) explicitly point out that the cognitive structure of emotion can in principle be highly dynamic and variable, both between individuals and over time. Although the dynamic and social properties of cognitions are investigated by sociologists and social psychologists alike, research has largely disregarded their links to emotion elicitation.

Given that cognitive structures for the most part arise in specific social contexts and relationships and—as outlined in Chapter 1—constitute distinct "thought communities" (Zerubavel 1997), it is reasonable to assume that individuals belonging to the same social units (i.e., are embedded in similar social structural configurations and systems of social order) appraise events in similar ways and therefore are also likely to converge in their emotional reactions.

Social cognitive processes of emotion elicitation

Research on the structural components of appraisal for some time tended to neglect the information processing architecture underlying appraisals and emotions. It seems clear that basic appraisals (such as evaluations of novelty) can be processed automatically and non-consciously, whereas more complex appraisals (such as norm compatibility) often require conscious and controlled processing. But how is it that appraisals based on seemingly complex and culturally highly variable meanings can rapidly and involuntarily elicit specific affective reactions without much conscious involvement or control? How is it that eating maggots, for example, triggers disgust in Westerners as quickly as the smell of rotten eggs? How is it that refusal to tip for good service can trigger anger just as instantaneously as a physical insult? And how can it be that losing one's job gives rise to fear just as automatically as the sight of a dangerous animal?

By taking account of such rapid and automatic appraisals, cognitive process theories have attempted—among other things—to do justice to some of the neuroscientific evidence on the non-conscious and rapid "low road" processing of emotion outlined above (Bargh & Ferguson 2000; Barrett, Ochsner, & Gross, 2007; Berridge & Winkielman 2003; Öhman *et al.* 2000; Robinson 1998). In contrast to earlier debates, they usually do not consider such automatic processes simply in opposition to cognition, but rather aim at their integration into a comprehensive process-oriented appraisal framework. Virtually all process models start from the assumption that appraisals are based on cognitive information processing in the broadest sense. These include basic sensory processing as well as higher order symbolic and conceptual processing capable of manipulating representations of objects, events, agents or situations on the basis of experience or newly acquired information (Reisenzein 2000). Reisenzein (2001, p. 190 f.) argues that process-oriented appraisal models have to address three issues.

First, they need to specify one or more representational media or formats in which the appraised events and the appraisals themselves are represented. Representational media may be fundamentally different in nature, for example iconic, auditory, or propositional. Furthermore, in view of the requirements to manipulate these representations, theories must include propositions about an underlying information processing architecture. This would concern, for example, assumptions about the storage, organization, manipulation, and retrieval of information. The specification of representational formats is particularly necessary when appraisals involve processes in which appraisal input and output use different representational media.

Second, the theories should specify the mechanisms of symbol manipulation (routines, algorithms, procedures) converting appraisal input or input representations into output representations. These specifications should deal, among other things, with the following questions: Are the operations that produce appraisal output from input representations innate or acquired, or are both types of processing conceivable? What additional information over and above the representation of a stimulus do these processes require to appraise an event, and how is this additional information represented? And under which conditions are these processes initiated (Reisenzein 2001, p. 191)?

Third, relationships between various appraisal processes must be specified in detail and process theories need to answer the question of whether different appraisal processes are operating constantly and whether they run sequentially or in parallel (Reisenzein 2001).

Sensory–motor, schematic, and conceptual processing

One of the earliest approaches addressing such issues is Leventhal and Scherer's (1987) hierarchical information processing model. Their account assumes three different levels of complexity in processing SECs: sensory–motor, schematic, and conceptual processing. According to this theory, different appraisal dimensions can be processed on each of the three levels and in corresponding representational formats (Leventhal & Scherer 1987, p. 17). Sensory–motor appraisal can be compared to the basic functions of the affect system. It "consists of multiple components, including a set of innate expressive-motor programmes and cerebral activating systems which are stimulated automatically, i.e. without volitional effort, by a variety of external stimuli and by internal changes of state" (Leventhal & Scherer 1987, p. 8). Schematic appraisals are learned and acquired during socialization and are later retrieved and automatically activated. Schematic processing combines:

> sensory–motor processes with image-like prototypes of emotional situations. Schemas are created in emotional encounters with the environment and are conceptualised as memories of emotional experience: They are concrete representations in memory of specific perceptual, motor (expressive, approach-avoidance tendencies and autonomic reactions), and subjective feelings each of which were components of the reaction during specific emotional episodes.
>
> (Leventhal & Scherer 1987, p. 10)

The conceptual level comprises reflexive, abstract, and deliberative processing of information on the basis of declarative knowledge. It "activates propositionally organised memory structures which have been formed by comparisons over two or more emotional episodes. Conceptual processing is also volitional and can evoke emotions by accessing schemas" (Leventhal & Scherer 1987, p. 11).

More recent emotion theories have elaborated and extended this conception of hierarchical information processing which is summarized in Table 2.3. However, most of them are based on only two levels of processing and assume that appraisal processes may differ considerably from one another even within the same representational formats, for instance propositional structures (Reisenzein 2000). These theories make reference to two fundamental features of Leventhal and Scherer's model which are also crucial for the overall argument of this chapter: First, they argue that information processing can be differentiated with reference to the degree of rapidity and automaticity. Second, they hold that the emotion system has a basic learning capacity which is related to the mode of information processing.

Although theories building on Leventhal's and Scherer's approach differ from each other in detail, they unanimously propose two fundamentally different modes of information processing implemented in specialized neural networks: automatic, associative, schematic, and heuristic processing on the one hand, and deliberative, active, controlled, and intentional processing on the other hand (Clore & Ortony 2000; Reisenzein 2001; Smith & Kirby 2000, 2001). These two modes largely mirror what "dual-process" theories of

Table 2.3 Processing levels for SECs

	Novelty	Pleasantness	Goal/need conduciveness	Coping potential	Norm/self compatibility
Conceptual level	Expectations: cause/effect, probability estimates	Recalled, anticipated, or derived positive–negative evaluations	Conscious goals, plans	Problem-solving ability	Self ideal, moral evaluation
Schematic level	Familiarity: schemata matching	Learned preferences/ aversions	Acquired needs, motives	Body schemata	Self/social schemata
Sensorimotor level	Sudden, intense stimulation	Innate preferences/ aversions	Basic needs	Available energy	(Empathic adaptation?)

Source: Reproduced from Leventhal & Scherer (1987, p. 17). Reproduced by permission of Taylor and Francis Ltd.

information processing posit for various kinds of cognitive processes (Evans 2008; Sloman 1996).

Associative and deliberative appraisals

Smith and Kirby (2000, p. 91) distinguish between associative and deliberative processing in emotion elicitation. According to their view, "appraisal detectors" are the hypothetical mechanisms that are primarily involved in computing appraisals. They process information from three different sources: direct perceptual input, associatively activated or retrieved representations, and the contents of working memory. Associative appraisals are presumed to rely on rapid, automatic, and memory-based modes of information processing implemented in non-hierarchical associative networks (Anderson & Bower 1973). Perceived stimuli activate memory traces of previous events on the basis of sensory properties or conceptual relationships with existing representations, with minimal use of attentional resources. Activation of these memories may in turn activate corresponding appraisal results, which can also be stored in memory and be recognized in a situation at hand by appraisal detectors, leading to the elicitation of an emotion (Smith & Kirby 2000, p. 93).

Associative processing can draw on virtually any type of representation stored in memory, such as representations of sensations, noises, tastes, smells, images, or of abstract concepts. By way of priming and the rapid activation of links within associative memory networks (*spreading activation*), appraisals that are attributed to past experiences or are elements of those experiences can be automatically retrieved, provided that features of the current situation concur with stored representations. This way, not only affective reactions but also more complex emotions can be triggered directly and rapidly (Smith & Kirby 2000, p. 94 f.).

In contrast to associative processing, controlled and deliberative appraisal proceeds more slowly, is more resource intensive, and usually requires conscious attention. For these reasons, it is considered to be conceptually mediated (Smith & Kirby 2000, p. 95). Also, deliberative processing is assumed to be considerably more constructive. It not only demands memory resources but also makes active use of the contents of working memory, which are frequently manipulated in order to produce appraisals. Smith and Kirby (2000) further assume that deliberative appraisals have limited access to representational media and can primarily draw on representations that have previously been semantically encoded. Basic sensory information usually does not become part of deliberative processing unless it has been linked to meaningful concepts. Smith and Kirby define both associative and deliberative modes as exclusively cognitive processes (Smith & Kirby 2000, p. 99 f.). Although they do speculate that appraisal detectors are located in subcortical brain areas such as the limbic system, they do not account for the possibility that early activation of the basic affect system may initiate associative processing in the first place and, under certain circumstances, strongly influences it (Smith & Kirby 2000, p. 93).

Appraisal reinstatement

A comparable process model of emotion elicitation has been developed by Clore and Ortony (2000). This model is based on earlier works (Ortony *et al.* 1998) and adds to them a level of automatic processing to integrate rapid and automatic emotion elicitation. Clore and Ortony (2000) suggest the notion of "appraisal reinstatement," which refers to the retrieval and reinstatement of appraisals that have previously been computed and stored in memory—as opposed to actively computed appraisals. They assume two corresponding modes of information processing: "bottom-up" processing for the active computation of appraisals and "top-down" processing for appraisal reinstatement. When appraisals are reinstated, the properties of previous situations that gave rise to certain emotions are compared with perceptions of the current situation. If this comparison reveals sufficient overlap, then it is likely that similar emotions will arise in the current situation. This type of automatic appraisal is not inconsistent with a predominantly cognitive approach, a view supported by research on automaticity in social cognition and behavior (Bargh 1997; Bargh & Chartrand 1999). According to this view, appraisal processes can be considered cognitive as long as they endow stimuli with meaning and significance. Thus, the two basic types of appraisal processes can be conceived of as special instances of cognitive processes, which are as such not restricted to affect or emotion.

Much like Smith and Kirby's (2000) account, the concepts of the reinstatement and active computation of appraisals draw on well-established "dual-process" theories of information processing (Evans 2008; Sloman 1996). According to these theories, the two modes of appraisal involve case-based associative processing on the one hand, and rule-based processing on the other hand. Clore and Ortony (2000) suggest that rule-based processing need not take place, but can occur—similar to associative or case-based processing— outside of conscious awareness. They suggest that these two modes are linked to two mechanisms categorizing emotion and information: first, *prototypical* categorization, which is characteristic for associative case-based processing and, second, *theory-based* categorization, which focuses on the symbolic and conceptual aspects of stimuli rather on their sensory properties. Prototypical categorization is quick and error prone, while theory-based categorization is comparatively slow and resource intensive, but more accurate.

Two key functions of emotion are associated with the two modes of information processing: preparing the organism for rapid behavioral responses and actions and decoupling stimulus from response to ensure flexibility and adaptability (Scherer 1994). The apparent contradiction between these two functions is resolved, at least conceptually, in Clore and Ortony's account. The notion of associative appraisal largely tallies with neuroscience findings on the basic affect system, which indicate that affective reactions can indeed occur without conscious awareness of a stimulus (and as a component of an emotion proper) (LeDoux 1996; Öhman *et al.* 2000). However, affective reactions for the most part do not trigger rigid behavioral responses, but rather *impulses* to act (*action readiness*) (Frijda 2004). As a result, by way of learn-ing and conditioning, certain stimuli become associated not with specific behaviors but rather—as also suggested by Rolls (1999, 2002)—with behav-ioral tendencies. "Such protocognitive processes allowed behavior to be contingent on a stimulus, but not dictated by it" (Clore & Ortony 2000, p. 41).

Assumptions about the basic appraisal processes can be mapped onto the neural level and integrated with corresponding neuroscience findings to produce an overall picture that consists of subcortical associative processes and more flexible and deliberative cortical processes instigating more or less rapid and affect-laden impulses to act. However, as Clore and Ortony (2000, p. 42) note, the location of the processes is not a sufficient criterion to label a process as either cognitive or precognitive. They suggest that a stimulus has to be evaluated in (subcortical) structures of the affect system to generate an affective response. Yet, in order to evaluate a stimulus, some form of representation has to exist to which the current stimulus can be compared and be endowed with affective meaning. This attribution of affective meaning clearly is a cognitive process (Clore & Ortony 2000).

Central, peripheral, and schematic appraisals

In another process approach to appraisal, Reisenzein (2001) distinguishes "central" (core) from "peripheral" appraisals. Peripheral appraisals are further divided into schematic and non-schematic appraisals, while central processes cannot be schematic. The two core appraisal processes compute (a) the belief- and expectedness-congruence as well as (b) the desire-congruence of focal events. In other words, core appraisals compute the congruence of occurring events with existing beliefs as well as the degree of an event's expectedness and the overlap of an event with existing desires. Importantly, and in contrast to their propositional input representations, core appraisals generate non-propositional, non-conceptual (analogue) output and are assumed to be hardwired, at least in terms of their functions (Reisenzein 1998).

Core appraisals operate constantly, in parallel, and outside of conscious awareness; their objective is to monitor congruence between newly acquired beliefs, pre-existing beliefs, and an individual's desires. The results of these appraisal processes give an indication of the degree of expectedness and the desirability of an event. In particular, their function is to direct attention towards critical and meaningful input propositions and to trigger characteristic affective reactions in order to prepare the organism for adaptive behavioral responses.

Peripheral appraisal processes, on the other hand, are propositional inferences that produce beliefs much in the same way other cognitive processes do. Peripheral appraisals determine the rigidity and strength of beliefs in view of an appraised event as well as its desirability. Furthermore, they generate propositional knowledge about the causes of an event, its possible consequences, and conformity to norms (Reisenzein 2001, p. 194). Thus, peripheral appraisals deliver most of the direct input to the core appraisal processes.

For example, certain beliefs about September 11, 2001, would be the outcome of peripheral appraisal processes based on the available knowledge and inferences of causality, norm compatibility, and the possible consequences of the event. These beliefs act as input to the core appraisal processes, through which they are compared with pre-existing beliefs, desires, and intentions. Central appraisals would categorize this event as virtually unprecedented and in conflict with existing beliefs and desires, thereby giving rise to corresponding emotions. On the other hand, a person sympathetic to the perpetrators of the attack would produce similar results with regard to peripheral appraisals, but central appraisal processes would produce a strikingly different output.

Reisenzein (2001) divides peripheral appraisals into schematic and non-schematic processes. Non-schematic appraisals are actively computed propositional inferences relying on general-purpose or domain-specific procedures in view of the appraised event. As such, they are comparable to "epistemic" and goal-directed actions operating sequentially, consciously, and in a resource-intensive way. In view of the assumed propositional medium of representation, schematic processing requires that schemas are structured representations made up of the basic elements of the medium of representation. A propositional medium is not dissimilar to language and can thus be linked to a "language of thought" with a corresponding semantic structure (Reisenzein 2001, p. 193). In Reisenzein's model, schemas influence appraisals in three ways (p. 193): (1) They reflect an actor's background knowledge which provides the basis for the initial conceptualization of an emotion eliciting event that constitutes the input for an appraisal process; (2) Schemas contain "general statements" and in conjunction with additional information allow establishing the premises from which appraisals of events can be computed; (3) The results of appraisal processes (apart from those of the two core appraisal processes) can be stored in memory from which they can be rapidly retrieved. In this way, appraisals can become components of schemas.

Importantly, Reisenzein (2001) emphasizes that both schematic and non-schematic appraisals are based on interactions with the social environment. Although—and in contrast to core appraisals—peripheral appraisals begin and terminate as mental states, they cannot be regarded as entirely internal to an actor. One example would be an inference that takes as its premise an input representation, additional concurrent information, and stored information from general knowledge structures and then generates the propositional content of an appraisal as its result. Sticking to the example of 9/11, it can be assumed that in peripheral appraisals of the event actors fall back not only on sensory perceptions but also on associated schemas and corresponding meanings (for instance explosions, catastrophes, hijackings, terrorism), the appraisals of other agents (for example media representatives, politicians, academics) or other contextually relevant situations (the Middle East conflict, the second Gulf War, etc.). Information contributing to the appraisal process in this case is also provided by the social environment and other actors and follows the logic of social and socially shared cognitions outlined at the beginning.

Reisenzein (2001) places considerable emphasis on this social and interactive aspect also of non-schematic appraisals, since most theories tend to view individuals as "isolated and nonsocial information processors, whose contact with the environment is restricted to the initial pickup of information about an initiating event on the input side, and the resulting emotional reaction on the output side" (p. 196). The link which appraisals establish to society and culture is an important component in Reisenzein's model and connections to the social environment are not only present in the form of general knowledge structures (or schemas), but also during the actual appraisal process, whether through the acquisition of additional information or to solicit support of other agents in making appraisals. In this way, Reisenzein (2001) creates an innovative perspective on the transmission and constitution of appraisals between actors and in social interactions.

An integrative view on appraisal processes

In view of these process theories of appraisal, which are summarized in Table 2.4, and the neuroscientific evidence on emotion generation outlined above, it seems that neuroscience and appraisal theories are mutually complementary rather than opposing one another (Parrott & Schulkin 1993; Scherer 1993a). This is particularly relevant regarding explanations of the social structuration of emotion and will be illustrated in the following attempt at an integrative view.

Table 2.4 Overview of process theories of appraisal

	Ortony & Clore (2000)		Reisenzein (2001)			Smith & Kirby (2001)	
Appraisal	Bottom-up	Top-down	Central	Peripheral active	Peripheral passive	Deliberative	Associative
Input	Theory-based	Prototypical	Propositional	Propositional	Propositional	Propositional	Any
Processing	Active computation	Reinstatement	Innate	Active computation	Passive recall	Active computation	Spreading activation

Basic affective reactions can be triggered rapidly, automatically, and non-consciously and give rise to a number of reactions that constitute defining features of an emotion proper, for example physiological arousal and action tendencies. Rapid information processing, in particular in subcortical areas of the affect system, and the quick initiation of physiological responses are components of an emotion that are as fundamental as, for example, a phenomenal feeling. In view of rapid subcortical processing, this raises the question of whether and how representations of stimuli can be stored in these networks and associated with affective valence. More often than not, the meaning of a stimulus is initially determined by subjective experience and not through symbolic communication or cultural learning. Hence, stimulus–affect associations and corresponding representations have to be made accessible to rapid "low road" processing. Even in cases in which the meaning of certain properties of stimuli (as proposed, for example, by Damasio (1994)) is hard-wired, some medium of representation and memory still is required. This suggests that at least some anatomical structures of the "low road" need to be capable of representing stimuli and comparing them with currently available sensory input. It thus appears that even the basic affect system performs operations that are comparable to appraisals, as is the case, for instance, at the level of sensory–motor processing in Scherer's model (1993a).

A typical characteristic of some automatic appraisals, however, is the assumption that they are based on schemas and associative networks of (propositional) representations, which usually contain models of situations that either encompass actual appraisal results computed previously or rapidly make available situation-specific and schema-consistent information for automatic appraisals. In this respect, neuroscience models differ from appraisal theories primarily in their conceptualization of appraisal input. In some neuroscience models, such as Le Doux's "low road" concept, only rudimentary sensory information serves as input to the emotion elicitation process (see also Rolls 1999, p. 104 f.).

This is accounted for by Scherer's (1984, 1993a) level of sensory–motor processing and by Smith and Kirby (2000, p. 93) who consider non-propositional representations as appraisal input in associative processing, which they locate at subcortical levels. Reisenzein's (2001) model, on the other hand, only allows for propositional input to central and peripheral appraisals, restricting non-propositional representations to appraisal outputs. This restriction is hardly justifiable, since it foregoes explanatory potential of appraisals in subcortical structures based on, for example, image-like representations or body-schemas. If non-propositional representations are accepted as appraisal input, it seems clear that there are two basic possibilities for associative and automatic appraisals: "Subcortical" appraisals which are based on simple and rudimentary non-propositional representations, and "schematic" appraisals drawing on propositional schemas embedded in associative and sensory cortical networks, such as those emphasized by Rolls (1999) (see also Smith & Kirby 2000; Squire 2004).

If we further assume that propositional schemas cannot be stored in subcortical networks and are more likely to be processed in cortical brain areas, this warrants the question whether and how these levels of processing interact with one another and under which conditions propositional input representations are processed either automatically or deliberately. Clore and Ortony (2000, p. 42 f.) assume that associative automatic appraisals are probably also computed in subcortical brain areas. If this is indeed the case, then an unresolved problem relates to the limited representational capacities of these brain areas, which are most probably restricted to non-propositional representations. It therefore seems unlikely that propositional schemas can be retrieved and processed in subcortical areas. If we further assume that appraisals of sensory perceptions are initially processed in subcortical, limbic brain structures and only subsequently undergo more differentiated processing in cortical regions, this suggests that earlier subcortical appraisals have a decisive influence on subsequent cortical processing or at least on the selection of the mode of information processing in these areas (deliberative vs. associative).

Thus there is evidence suggesting that subcortical processing has one key advantage over cortical processing, namely that it is considerably quicker (LeDoux 1996). This does not only mean that individuals can react faster to possible threats, but also that cortical processing may under certain circumstances be influenced by the outcomes of rapid subcortical processing, since it kicks in chronologically later. This is extensively reflected by studies on the influence of affect and emotion on other forms of cognitive information processing, such as reasoning, judgment, and decision-making.

Figure 2.3 outlines this integrative view and illustrates the interplay of associative and deliberative appraisals in the process of emotion elicitation. External events are initially subject to perceptual processing whose outcomes are then categorized by associative processing and possibly linked to pre-existing cognitive schemas. Subsequent automatic appraisals—in subcortical areas based on rudimentary representations, in cortical areas on propositional schemas—give rise to an affective response and corresponding physiological and psychological reactions. In parallel, and largely outside conscious awareness, sensory processing gives rise to somatosensory (physiological) sensations, for instance an unpleasant pressure on the skin, which are then consciously recognized, conceptualized, and categorized, for instance as a jostle. This recognition provides the basis for the active and deliberative appraisal of an event, during which available contextual information is evaluated. If the jostle initially produces affective reactions such as anger or annoyance, active appraisal may well lead to the inference that it happened accidentally. True, active appraisal also accounts for the basic, initial affective reaction, but may ultimately give rise to an emotion other than anger or annoyance. Importantly, the earlier basic affective reaction also influences active appraisal processing, not only as a component thereof, but also by originally influencing somatosensory sensation and recognition stages (see Chapter 3).

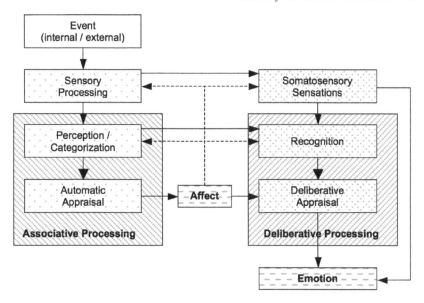

Figure 2.3 Associative and deliberative appraisals.

Consequently, dual-process assumptions and the "two systems of reasoning" (Sloman 1996) underlying process theories of appraisal are addressed and influenced by a system that may be characterized as "cognitive," in the sense that it processes and manipulates information, but which operates on non-propositional representations in the subcortical affect circuits of the brain. One might thus be on safer grounds by not speaking of "reasoning systems" but instead of "affect systems" (see also Duncan & Barrett 2007). It should also be noted that deliberative (non-schematic peripheral) appraisals are similar to "conventional" cognitive inferences and therefore subject to the influence of earlier affective reactions (see Bless 2000, p. 202; Clore & Huntsinger 2007; Clore & Ortony 2000, p. 27).

In summary, appraisal theories offer a well-developed framework for explaining emotion elicitation that is certainly focused on the level of the individual, but is also fundamentally capable of accounting for the manifold social influences on emotion. At the same time, their potential lies in simultaneously accounting for the structures and processes of emotion elicitation. They therefore constitute an important contribution to any sociological approach aimed at uncovering the social structuration of emotion. Based on the integrative view outlined above, the following sections will further address the key question of this chapter, namely how social structures and social order systematically influence the various levels of emotion elicitation.

The social structuration of affect and emotion

In the two previous sections, I have discussed the neural and cognitive bases of the generation of basic affects and complex emotions proper. I placed considerable emphasis on automatic and non-conscious as well as on controlled and deliberative processing and suggested that the mechanisms and processes underlying these types of processing are, at least partly, susceptible to social environmental shaping. Without the ability to account for the—stable as well as dynamic—aspects of the social and cultural environment, emotions would be virtually indistinguishable from instinct-like stimulus–response behavior. According to some views, emotions primarily evolved to "decouple stimulus and response" (Scherer 1994). Rolls (1999) makes a similar claim arguing that stimulus–reinforcer learning, although comparable to operant conditioning, does not automatically prompt behavioral responses, but rather instills affective reactions, which are considerably more flexible in guiding behavior. Thus, from a sociological perspective, one outstanding property of emotions is that they allow linking more traditional sociological determinants of action (such as norms and rationality) to automatic and rapid forms of behavior in ways that have previously been discredited as being biologically reductionist. Therefore, this section will examine in detail the question of how the structural and symbolic orders in which agents are embedded are reflected in the elicitation of emotion. To do so, I will first illustrate the experience-dependent plasticity of the biological architecture of the brain's affect systems. Second, I will highlight the role of implicit emotional memories in emotional experience; and third emphasize the social origins of the contents and structures of appraisals.

Experience-dependent brain plasticity

Sociological emotion theories have been criticized for their neglect of the biological bases of emotion (Turner 2000), a critique that is often also applied to other areas of sociological inquiry (e.g., Benton 1991; Freese *et al.* 2003; Newton 2003). This almost traditional disdain of biological and physiological processes is based on the assumption that they are some of the organism's immutable reference points, which makes them unsuitable for inclusion into sociological arguments. However, research increasingly shows that the opposite is in fact the case. In many areas of sociological inquiry, it is more and more recognized that the social environment shapes individuals not only in terms of cognitions and actions, but also on the bodily, physiological level. This kind of "social constructionism" emphasizes that not only is knowledge socially shaped—as Berger and Luckmann (1969) outlined in their seminal treatise—but also the very biological architecture on which knowledge, cognition, and action are realized. Thus, an intriguing question is whether the social shaping and structuration of emotion can also be located and conceptualized at the level of human physiology?

A number of social influences on the information processing architecture underlying affect and emotion can be identified on the level of biological brain development, commonly referred to as brain plasticity or neural plasticity. For quite some time, social influences on brain architecture, as marshaled by the "social brain hypothesis" (Dunbar 2002), primarily focused on phylogenetic factors in brain development. Work in this tradition argued, for example, that the size and ontogenetically relatively late maturing of the human brain were primarily a consequence of adaptation to living in groups and larger social units and the increased demands for communication and cooperation (Eisenberg 1995). More recently, conceptualizations of the "social brain" have increasingly attended to ontogenetic factors and the effects of experience-dependent development on human neurophysiology (Brothers 1997; Lieberman 2007).

A number of studies have shown that ontogenetic influences play a decisive role in the stabilization and reorganization of the neural circuitry initially present in the human brain (e.g., Elbert *et al.* 2001; Kolb & Whishaw 1998). This means that biological brain maturation, both structurally and functionally, is supposed to depend to a large extent on the social environment, in particular social relationships—not just during primary socialization but throughout the entire life course, although plasticity decreases considerably with age (Braun 2011; Cicchetti & Curtis 2006). Thus, psycho-social influences, particularly during the early years of life, can lead to far-reaching alterations in neural networks, which in later life are difficult to change or reverse, and it has been argued that limbic affect systems are particularly receptive to these influences (Braun 2011; Davidson *et al.* 2000).

Although current research still strongly relies on animal studies or focuses on cellular and molecular levels, looking at, for example, dendritic length, synaptic connectivity, and neurotransmitter and metabolic brain activity, some studies have shed initial light on the significance of these changes at the human behavioral level (Davidson *et al.* 2000, p. 900 f.; Kolb & Whishaw 1998). In a review of existing research, Kolb and Whishaw (1998) argue that experience-dependent brain development does indeed give rise to a number of functionally independent changes in brain anatomy and that these changes are in fact behaviorally relevant. Initial evidence also indicates that the limbic system is particularly adaptive in terms of neural plasticity and that the quality of socio-emotional experiences and interpersonal ties (as opposed to "purely" cognitive activities) is one of the most influential forces in altering brain structure in these regions (Bock *et al.* 2003; Cicchetti & Curtis 2006). Caregiver–infant relationships in early years of life have proven to be crucial in neurogenesis (Cynader & Frost 1999; Davidson *et al.* 2000) and certain neural structures, whose role in emotion elicitation has been outlined in Chapter 2, are particularly susceptible to these influences (Bock *et al.* 2003, p. 53). Davidson and colleagues propose that associations between stimuli and affective responses, as described in earlier sections of this chapter, can be conceived of not only in representational or associative network terms, but also as changes at the molecular level (Davidson *et al.* 2000, p. 900).

Thus, it seems plausible that events and experiences occurring regularly within some social unit may give rise to comparable developments in brain structure across many individuals. One argument in favor of this view is that, to be "socially adequate," the developing brain requires stable social structures and socio-emotional ties consistently providing interactions and information needed for successful maturation, in both social and biological terms (Cynader & Frost 1999; Davidson *et al.* 2000). The disruption or deprivation of this input may cause failures in the adaptation of the neural circuitry and increase the likelihood of, for instance, learning and behavioral disorders or even serious mental illness in later life (see Bock *et al.* 2003). Most interestingly, recent research has also demonstrated notable influences of socio-economic status on brain development (Hackman & Farah 2009; Hackman *et al.* 2010). Similar to the well-established effects of socio-economic status on physical and mental health (Marmot 2004; Wilkinson & Pickett 2009), individuals' status decisively impacts a range of neurocognitive abilities, in particular those related to language and executive functions. Candidate mechanisms include prenatal development, parental care, and cognitive stimulation (Hackman *et al.* 2010; Noble *et al.* 2007). In the following section, however, I will shift the focus of analysis from alterations in biological brain architecture to a "representational" view, in which past emotional experiences are assumed to be stored in different memory systems of the brain which exhibit notable but distinct influences on appraisals at various levels.

Emotional memory

Looking for the possible sources of social influence on affective information processing, it is interesting to ask whether the susceptibility of cognitive appraisal structures to social shaping is in some way mirrored in the subcortical circuits of the basic affect system, in which representational capabilities and flexible learning and relearning are limited. This problem is much less severe in view of cortical areas, which are well known for their role in semantic and conceptual processing. The key challenge in uncovering the social shaping of emotion at the physiological level therefore relates to the subcortical affect system and prompts the question whether non-innate and possibly complex representations (of the social and cultural world) can be stored or accessed in subcortical circuits and contribute to imbuing stimuli with an affective valence based on previous experience.

One possible way of attending to this question is looking for traces of evidence on an adequate memory architecture and related representational media outside the prefrontal cortex. Since it has been shown that appraisals do not necessarily require propositional representations but can also work on non-propositional representations, a simple capacity to store percepts, physiological states, and corresponding behavioral responses may provide a sufficient basis for producing appraisals. It thus seems reasonable to examine the mechanisms of emotional memory and memories for emotions in more detail.

A number of case studies on individuals with damage to certain brain areas have been as revealing for memory research as they have been for emotion research. Standard psychological tests of memory performance are mostly based on the capacity to recall verbal, conceptual, and semantically encoded information. Naturally, assessments making use of affective and emotional stimuli produce very different results. In overcoming the limitations of existing tools assessing cognitive memory performance, studies using alternative measures and techniques have shown that situations involving various forms of reward or punishment (such as pleasure and pain), although they are often not consciously remembered, do reliably trigger affect-congruent approach or avoidance behavior. LeDoux (1996), for example, reports the case of a patient who was unable to form new declarative memories, for instance of events or persons. If, however, the events or persons involved strong affective reactions (such as pain), the patient tended to avoid those events on future occurrence— but was unable to give reasons for this avoidance. This means that, despite the lack of conscious memory for rewarding or punishing stimuli, situational factors associated with these stimuli can somehow be stored, retrieved, and converted into behavioral responses (LeDoux 1996, pp. 182 ff.; Welzer & Markowitsch 2001).

These and other findings have led to the now widely held assumption that there are (at least) two distinct memory systems that differ in view of their contents and the time spans they cover (short-term vs. long-term memory): a declarative, explicit memory system and an implicit, non-declarative system. Both are linked to various levels of consciousness involved in processing information and representations (Squire 2004). Only the contents of the explicit, declarative system can be "consciously retrieved, flexibly deployed, and combined with new information" (Welzer & Markowitsch 2001, p. 207; own translation). Declarative memory is made up of a system for facts (semantic memory) and for events (episodic memory). As Figure 2.4 shows, the implicit systems can further be divided into procedural and priming memory and systems for conditioning and non-associative learning (Squire 2004, p. 173).

Much of the sociology of knowledge deals with "stocks of knowledge" that are primarily components of declarative memory. The contents of this system (semantic and episodic memory) are represented predominantly in extensively ramified cortical networks (Squire 2004, p. 173; Welzer & Markowitsch 2001, p. 207), in which the hippocampus and amygdala, as subcortical structures, play a central role in the initial storage and consolidation of memory (McGaugh 2003; Phelps 2004). Besides autobiographical memories, which are stored in the episodic system, knowledge representation in these networks also encompasses conceptual representations in the form of propositional semantic networks and schemas.

It seems likely that in particular episodic autobiographical memory is closely linked to affect-related memories, which can in turn be represented

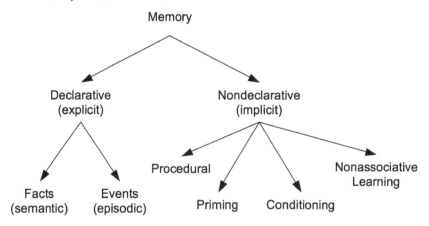

Figure 2.4 Taxonomy of long-term memory systems.
Source: Adapted from Squire (2004, p. 173).

in implicit structures. There is considerably less agreement, however, as to the location of these implicit, non-declarative (or habitual) memory systems, mostly because of the pronounced differences in this category. Squire and colleagues, for example, locate all non-declarative forms of memory—except for priming memory—in subcortical brain areas (striatum, amygdala, cerebellum, basal ganglia) (Bayley *et al*. 2003; Squire 2004). The amygdala seems to be of central importance here: It not only plays a key role in emotional learning and conditioning, as LeDoux (1996) has shown for fear conditioning, but evidently also modulates the strength of declarative and non-declarative memory (see Hamann 2001; LaBar & Cabeza 2006; Phelps 2004).

A key question therefore is whether non-declarative memory can actually be regarded as a representational form of information storage. Squire (2004, p. 173) argues that in contrast to declarative memory, which is representational and provides a model of the external world that is either true or false, non-declarative memory is neither true nor false but rather *dispositional* and expressed through *performance*, not recollection. Contents of non-declarative memory are therefore thought to be retrieved by reactivation of the systems in which learning initially occurred (p. 173).

Without further engaging with this question of representational media and capacities, it is well established that the subcortical structures of the implicit memory system are also implicated in the storage of information in declarative memory. With regard to emotion elicitation, LeDoux distinguishes "emotional memory" from the "memory of emotions" (LeDoux 1996, p. 182) (see Figure 2.5). The memory of emotions is based on the retrieval of declarative knowledge of specific situations and of the emotions experienced in these situations. Emotional memory, on the other hand, denotes the implicit memory of emotionally arousing events, i.e. those events that have triggered affective

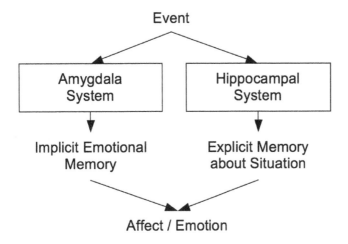

Figure 2.5 Emotional memory and memory of emotion.
Source: Adapted from LeDoux (1996, pp. 202 ff.).

reactions or emotions proper in the past. The difference between the two systems is that, generally, emotional memory recreates the physiological characteristics of a past situation. In other words, it induces a certain physiological state and thereby plays a decisive role in the replication of the phenomenal sensation, the subjective feeling experienced in a situation. Declarative memory, on the other hand, provides factual and contextual information of the past situation. If I find myself, for example, at a certain place and feel ill at ease, but without knowing exactly what the cause of this unease might be, the feeling may be attributable to an emotional memory. If I subsequently remember an event that took place at this particular place and the emotions I experienced at that time (which presumably are the cause of my present unease), I am remembering emotions via declarative memory.

Thus, emotional memory triggers the relevant emotion or, in the first instance, its basic affective components, before additional declarative knowledge about the situation in question is retrieved to further specify the affective reaction with conceptual (emotion) knowledge (LeDoux 1996, p. 201). This mode of emotion elicitation, based on past experiences, requires stored models of situations and events that have to be combined with affective reactions and which, once combined in this way, could also be described as stored (schematic) appraisals. As part of implicit memory, the amygdala-based memory system can be regarded as the counterpart of semantic memories and serves as the basis for automatic, case-based associative appraisals.

In essence, Damasio's (1994) somatic marker hypothesis precisely relates to this phenomenon. Somatic markers are considered associations of certain courses of action and their emotional consequences that have occurred in the past. If an event with corresponding behavioral options occurs and corresponds to stored event models, then the appraisals and affective reactions stored as components of this event schema are triggered, highlighting certain options for action. Somatic markers can thus be conceived of as specific instances of emotional memory (Damasio 2003, p. 147 f.; discussed in more detail in Chapter 3). According to Damasio, the physiological states and reactions that are characteristic of emotions and to which somatic markers point are represented in somatosensory cortices. The representations of physiological arousal can be activated—and phenomenally felt—simply on the basis of appropriate signals from the amygdala (or, more slowly, from prefrontal areas)—without actually arousing the physiological state. Damasio (1994) calls these phenomenal experiences "as-if-feelings" (p. 159). Thus, the non-conscious and automatic retrieval of stored appraisals and patterns of physiological arousal do reflect actors' experiences and are a focal area of emotional socialization.

This implicit emotional memory system contains traces and images of past events, situations, and actions, together with the emotions that accompanied them. Given that actors are socialized under conditions of more or less stable social orders (both symbolically and structurally), which are characterized by recurring patterns of social interactions, embeddedness into social institutions, and stable sactions and reward contingencies, then it seems reasonable to assume that large numbers of actors within a social unit share comparable implicit (and explicit) memories and stocks of knowledge with comparable emotional connotations.

Thus, one of the effects of the highly institutionalized life courses and biographies in modern societies is that virtually every individual has specific experiences with, for example, authorities (e.g., teachers, parents, officials), collective actors (the state, bureaucracy), and other institutions (the family, educational system, clubs) that undoubtedly differ from each other in detail, but tend to be similar in their general structure and focus and to be socially shared within different groups (e.g., social classes). Paralleling the arguments of the sociology of knowledge outlined at the beginning, this suggests that implicit emotional memory is structured in similar ways as declarative knowledge. Consequently, social reality determines not only what we think and do, but also what we sense and feel. Similar to the accounts given by Mannheim (1936) and Berger and Luckmann (1969), society produces not only certain (cognitive) knowledge structures, but also affective structures which may even be the basis for the former.

The social structures of emotion postulated in this section differ from those proposed in many variants of emotion sociology in that the latter often assume that deliberation, reflexive thought, and symbolic meaning are the main pathways through which emotions are shaped by culture and society. The view presented here may supplement existing social structural accounts by emphasizing that the social shaping of emotion is not accomplished exclusively by the downward influence of higher order cognitive processes, but rather in a complex interplay with basic affective reactions. This means that the social structures of emotion in fact also encompass automatic and core affective reactions that are triggered largely outside conscious awareness but nonetheless exert a decisive influence on behavior (see Chapter 3 for details).

Opposing this view, it could be argued that implicit memory for the most part depends on personal and subjective experience and corresponding physiological reactions and emotions. However, in modern and functionally differentiated "knowledge societies," in which a shift away from physical and subjective experience towards mediated experiences and declarative knowledge has been diagnosed, the implicit memory system would be of considerably less significance for the argument advanced here. This is because an increase in factual knowledge at the expense of first-hand experiential knowledge might also decrease the degree of socially shared implicit emotional memories, thus weakening the assumption of the social structuration also of basic affective reactions.

This objection, however, can be countered by looking at evidence from studies indicating that declarative memories represented in cortical networks, created by verbal learning or explicit acquisition without direct subjective experience, do activate specific amygdala responses characteristic of experience-based emotional memories upon retrieval. Generally, the formation of implicit experiential memories is accompanied and in fact facilitated by a corresponding activation of the affect system, whereas verbally mediated declarative memories may be formed without significant affective connotation. Interestingly, Phelps and colleagues (2001) have demonstrated that the amygdala is also activated by purely "cognitive" representations of fear and not only when fear is or has actually been experienced. These findings support the view that initial processing via the "high road" involving propositional representations can also give rise to characteristic patterns of activation in the basic affect system, in particular the amygdala, similar to activation in response to direct sensory stimulation (Damasio *et al.* 2000).

Consequently, we can assume that implicit and experiential emotional memory is at least as important as (or even subserves) declarative memory and explicit knowledge in terms of social action and the navigation in the social world. As input to appraisal processes, implicit, habitual memories are important automatic and pre-reflexive triggers of emotions. Importantly, conscious attention towards a stimulus is neither necessary for storing nor for retrieving implicit memories (Morris *et al.* 1998). It is hardly surprising then that peripheral contextual information stored in implicit memory, together with the actual event that originally triggered the emotion, may under certain circumstances be sufficient to trigger an emotional response. From a sociological point of view, this is all the more important since the mere association of a stimulus with a particular category or its classification even as a signifier of a particular social category or an element of social distinction (e.g., in terms of social class, status, power, ethnicity, nationality) may trigger affects and emotions that have previously been triggered by or associated with a specific (but possibly different) element of such a category (see Olsson *et al.* 2005).

Naturally, automatic emotion elicitation is not the only pathway through which the social structuration of emotion can be reconstructed, but it is one that is best suited to explain the ways in which emotional action contributes to the reproduction of social order and social structure, thereby highlighting the dual nature of micro–macro linkage. Compared to the deeply embodied implicit and non-conscious elements of emotion elicitation, explicit and symbolic knowledge—though important in its own respect—is more susceptible to discursive processes, relearning, and revision and therefore potentially less stable and persistent. The social shaping of affective responses may thus be more rigid and stable at implicit levels (as is evident from pathological cases, such as phobias), while (shared) declarative memories seem more fragile and more susceptible to change (cf. LeDoux 1996, pp. 225 ff.; Markowitsch 1999).

Emotional memories as well as somatic markers anchor individuals in their lifeworlds through the affective meanings manifest in previous experiences. Implicit memories systematically influence ongoing appraisals via information associated with immediate situations, even if an individual is unaware of this information. This may not only be "cold" information such as contextual cues, but also "hot" information in the form of somatic markers (patterns of physiological arousal) that already contain specific "affect signatures" of previous appraisals. Certain moods, such as uneasiness, whose causes are often hard to establish, are examples at hand. Further examples are certain (impulsive) behaviors or the intuitive liking or disliking of others that we can often hardly justify or explain.

Thus, during socialization, individuals develop robust and effective "affective dispositions" or "fundamental sentiments" (Davidson 2003b; Heise 1979; Lazarus 1991a; Lazarus & Smith 1988)[9] in view of various categories of events, acts, and objects. These include not only specific, narrowly defined stimuli but also all everyday objects and phenomena, such as consumer goods (Erk *et al.* 2002; Yeung & Wyer 2004), signs and symbols of social distinction (Mowrer *et al.* 2011; Zink *et al.* 2008), or resources such as money (Knutson & Bossaerts 2007; O'Doherty *et al.* 2001).

In view of the social structuration of emotion elicitation, it is reasonable to assume that these dispositions are all the more robust the more consistently and regularly elements of these categories occur and give rise to comparable emotional consequences, thereby strengthening existing dispositions. Repeated unpleasant experiences with authorities, for example, might reinforce aversive affective dispositions. Experiences with certain consumer goods deemed rewarding and desirable would shape the basic affective reason to such types of goods more generally. Depending on one's own education and achievement, symbols of distinction, such as academic degrees, titles, or positions, may lead to greater or lesser admiration, while money also gives rise to corresponding reactions, depending on how scarce it is or was in the past. These mechanisms are also crucial in enhancing our understanding of collective emotional phenomena, such as group emotions and emotional atmospheres which, since Durkheim's notion of collective effervescence (Durkheim 1915), have been regarded by many scholars as basic building blocks of larger social units (de Rivera 1992; Scheff 1997; von Scheve & Ismer 2013).

In the following section, I will further complement this perspective that is primarily based on implicit, non-conscious, and non-propositional processes with a more detailed investigation of the higher cognitive foundations of emotions and how they are shaped by culture and society. My objective then is to show how social structures and social order shape the "structures of thought," i.e. the cognitive structures and propositional representations that are of crucial importance to conscious, controlled, and rule-based appraisals.

Social cognition and social representations

The previous section argued that the propositional processing of appraisals can hardly take place independently of earlier "non-propositional" appraisals based on implicit memory and the output of the basic affect system. Despite this influence, semantic and propositional information processing and higher order cognitions are indispensable to the elicitation of complex emotions. Furthermore, the established role of higher cognitions in inferring meaning from symbolic information, in accounting for abstract contextual cues and complex social information (e.g., regarding social relationships, networks, coalitions, etc.), or in predicting future states of affairs through inferences and deduction, suggest taking a closer look at the social construction of this propositional dimension of appraisal.

In view of the structures and contents of appraisal discussed above, there is broad consensus that appraisals establish relationships between events and actors' concerns in the broadest sense. Appraisals thus reflect the relevance of events and trigger corresponding affects and emotions. The question of the mutual dependencies within this relationship is usually answered quite clearly: Appraisal theories generally assume that appraised events are dynamic in nature, meaning that they are attended to and become immediately relevant only for a rather short time (cf. Frijda & Zeelenberg 2001). Actors' concern structures (cognitive or biological), on the other hand, are assumed to be more or less stable.

In this section, I will argue that, from a sociological (and ontogenetic- or socialization-centered) point of view, concern structures should not be conceptualized as stable and unalterable. A dynamic perspective on these concern structures is not only necessary for explaining the feedback and feed-forward processes in the emergence and reproduction of social order, but is also in line with the basic assumptions of cognitive sociology and some appraisal theories (e.g., Lazarus & Smith 1988; Ortony *et al.* 1988; Reisenzein 2001). Lazarus and Smith (1988) have illustrated these links in the most sociologically relevant way by incorporating "general knowledge structures" into their account of appraisal and distinguishing them from the effects of contextual knowledge. General knowledge structures equal our understanding of cognitive structures outlined in Chapter 1 and consist of relatively stable beliefs, attitudes, and everyday theories, which emerge in interactions with culture and society. Contextual knowledge also plays an important part in appraisal processing. It is acquired quickly and at times rather automatically and is based on perceptual information available in a current situation. Thus, contextual knowledge is a form of knowledge that results from the *definition of the situation* (Lazarus & Smith 1988, p. 283 f.). To that extent, contextual knowledge reflects the dynamic aspects of appraisal. It takes account of the fact that most "external" emotion-inducing events are initially registered through sensory perception, while categorization and recognition of perceptual information are experience-dependent and influenced by social and affective factors. In fact, I will argue in more detail that general knowledge structures systematically influence contextual knowledge (and hence the definition of the situation) and that this process is guided by affective reactions (see Figure 2.3).

For obvious reasons, psychology in general has shown little interest in the long-term and social variability of cognitions and cognitive structures. However, social psychological research on social cognition and social representations has produced a number of theories and studies on the genesis and malleability of individuals' mental states in relation to culture and society, as proposed by classics in social psychology such as Vygotsky (1978) and Piaget (1954) as well as by sociologists such as Cooley (1909), Mannheim (1936), and Mead (1934) (see Howard 1995, p. 91). Social cognition refers not only to the perception and processing of explicitly *social* information but also to the theories, images, and representations that actors produce of other actors, social relations, and culture and society more generally. Accordingly, cognition is usually prefixed as "social" when:

a.   its objects are social in nature (for example, other agents);
b.   it is social in origin (thinking about social hierarchies); or
c.   it is socially shared and distributed (used by various members of a society in a similar way).

(Leyens & Dardenne 1996)

Social cognition does not constitute a homogenous theoretical approach but usually a compendium of approaches and concepts used to investigate well-known phenomena such as stereotypes, attitudes, perceptions of others, and social interactions that basically explain, "how people make sense of other people and themselves" (Fiske & Taylor 1984, p. 17). Thus social cognition oscillates between the "construction of social reality" (Bless *et al.* 2004; Searle 1995) and the "social construction of reality" (Berger & Luckmann 1969; Moscovici 1961).

Sociology is indispensable when making predictions on the dynamics and social patterns of social cognitions and, accordingly, about the emotions based on or related to these cognitions. From this point of view, social cognitions are of particular importance to the aims of the present investigation primarily because they give precise accounts of actor's mental social embeddedness. They contain both autobiographical experiences in the social environment and shared knowledge of others and of society more generally, for example as knowledge of social institutions, groups, cultures, practices, values, conventions, and norms. The position adopted here is that sociality, which is mediated and experienced in the social realm (e.g., in social interactions or through various symbol systems), is internalized primarily through cognitions, i.e. initially through perceptions and subsequently through processes of categorization, classification, and schematization that generate enduring cognitive structures and principles of meaning-making (see Berger & Luckmann 1969; Cerulo 2002; D'Andrade 1981; DiMaggio 1997, 2002; Shore 1996; Zerubavel 1997). In this view, it seems plausible to conceptualize appraisal processes as genuinely "social" and also to conceive of the structural components of appraisal as socially constructed, in view of both the events that are appraised and the cognitive structures providing the basis for appraisal.

Social and socially shared appraisals

In explaining the social basis of appraisal, Manstead and Fischer (2001) give the example of anxiousness about exams, which may depend essentially on how fellow students appraise the forthcoming ordeal. Thus a person may be anxious about an exam simply because of others' anxiety, even though he or she does not regard the exam as particularly challenging. However, other actors' anxiety can serve as (additional) input to individual appraisal processes (Manstead & Fischer 2001, p. 221 f.). Another example concerns the reliability or appropriateness of one's own appraisals. Whether one finds a film amusing or boring, sophisticated or terrific, depends not only on an appraisal of the film itself but also on the reactions of third parties. In processing appraisals, agents refer to external appraisals not just in order to conform to norms or notions of what is socially appropriate, but also when unable to appraise an event, for example because of incomplete or too complex information or lack of cognitive resources. In these cases, appraisals are processed in a socially shared fashion, much like other forms of socially shared cognition (see Hutchins 1996; Oatley 2000; Resnick 1991).

Similar forms of shared cognition also occur when defining a situation, for example when a situation cannot be assessed by an individual alone, but has to be construed in interaction with others, for instance in negotiation or collaboration. Depending on the social relationships between actors, appraisals of others will be taken into account to varying degrees (Smith *et al.* 2006). This social interaction carries all the more weight since it is not tied to physical co-presence but can be distributed over time and space. Actors therefore can refer to appraisals that took place in the past or—in all likelihood—will take place in the future, and the referenced appraising entities do not even have to be individual social actors, but a collective or corporate body, a fictional character, or simply the "generalized other" in monitoring, for example, the congruence of own appraisals with social norms and conventions. According to this view, emotion norms (Thoits 2004) can also be seen as "appraisal rules," since they implicitly prescribe how actors are supposed to assess certain events. In this way, appraisal tendencies or patterns that are characteristic for a social unit can be acquired through learning (even institutionalized education) and socialization.

Importantly, these:

> precomputed appraisals can be communicated not only in the course of an encounter with a specific eliciting event; they can also be acquired from others as parts of schemas for events of the same or similar types long before a concrete instantiating event is encountered ... People undoubtedly acquire numerous schemas with stored appraisal information during their socialization in a culture.
>
> (Reisenzein 2001, p. 197)

In modern societies, a crucial role can be attributed to the mass media, since they disseminate information on how others have appraised or intend to appraise almost any event of international and domestic relevance (Buck & Powers 2011; Döveling & Schwarz 2010). Appraisals conveyed through the media may be of marginal impact in those situations in which a broadcast is actually perceived. However, having in mind the mechanisms of emotion elicitation reviewed above, it is conceivable that mass-medially transmitted appraisals are (non-consciously) retrieved in later situations and applied to an ongoing appraisal process, for instance as part of an event schema. But the mass media is only one pathway of appraisal transmission, which can happen through various channels of cultural transmission, for example, via institutions (in particular education), artworks, or interpersonal, interfamily, or intergenerational transmission.

Support for these assumptions is indeed found in the notion of socially shared or distributed cognitions, which refers primarily to expert knowledge and its application to specific types of problems and situations requiring cooperation, but which could also in principle be applied to appraisal problems. Wertsch (1991) and Resnick (1991) assume not only that cognitions are socially situated and constructed, but also that a prerequisite for cooperation is a sufficiently large overlap in people's stocks of knowledge. It has already been shown that such overlaps are constituted most generally under comparable socialization conditions; but have to be established by other means in specific domains (Cannon-Bowers & Salas 2001, p. 196; Hutchins 1991, 1996). A further possible source of influence emanating from the social sphere is constituted by appraisals of individuals, social actions, and *social facts*, such as roles, status, authority, or social relationships. A theory that is well elaborated in this regard but scarcely perceived as an appraisal theory is Kemper's work (1978), which focuses explicitly on appraisals of status and power.

In specifying the idea of social appraisal, Manstead and Fischer (2001) note that:

> whether or not something is perceived as frightening is largely dependent on how such a threat has previously been appraised and talked about by one's parents or peers. The assessment of our ability to cope with difficulties or threats is thus likely to be influenced by other people . . . Appraisals, in other words, are importantly shaped by the appraisals of important others in the same or similar emotional events.
>
> (p. 222)

They thus define social appraisals as appraisals of thoughts, emotions, and behaviors of other actors in view of an emotional event.

One of the key objectives and strengths of appraisal theory is explaining individual differences in emotional reactions to identical events. Sociology is interested in uncovering societal patterns in interindividual differences and similarities, as in research on social differentiation and inequality. This is precisely why patterns of socially determined interindividual differences in appraisal (and thus in resulting emotions) are crucial to sociology. Sociology may take as self-evident what is usually not the focus of attention in (social) psychology, namely that some of the cognitive foundations of appraisal are constantly emerging from social contexts and are accordingly shaped by culture and society: They are socially constructed.

Schematic appraisals and appraisal schematization

In appraisal theory, sociality is often only accounted for by those appraisal dimensions that are obviously of social origin, such as conformity to norms and values. Social norms, for example, reflect not only individual but also collective goals and beliefs as behavioral expectations (Smith *et al.* 2006). However, taking Scherer's (1984) SECs as an example, it can be argued in fact that most of these checks are susceptible to social shaping in different ways. Whereas novelty and intrinsic pleasantness might be affected through previous subjective experience, frequent exposition to similar events, and habituation, goal congruence, coping potential, and norm compatibility are more dependent on semantic and declarative knowledge and may involve still other processes.

One well-established principle of organizing these appraisal-relevant forms of knowledge is, again, schemas. As cognitive structures, schemas, which are made up of conceptual or generalized stocks of knowledge and certain expectations, are a possible link between the structural and processual components of appraisals. Leventhal and Scherer (1987) hint at this link with the level of schematic information processing. Schematic processing is also important to associative and case-based reasoning in appraisal (Clore & Ortony 2000; Smith & Kirby 2001). Schemas are also critical in understanding how initially deliberative and conscious appraisals based on complex information are transformed into non-conscious and automatic appraisals, whose properties are comparable to more rudimentary appraisals of the basic affect system.

Since schematic information processing is crucial to many forms of appraisal, looking at their emergence and operation may provide additional insights into the ways in which culture and society shape appraisal. Schema theory unveils the social dimension of cognition since schemas often emerge in the process of reducing the complexities of the social world. In schema theory, humans are seen as "cognitive misers" who aim at operating as efficiently as possible with limited (cognitive) resources. Consequently, schemas reduce the complexity of the social environment in and upon which they act and "automatize" frequently recurring acts of interpreting and assigning meaning to events (Fiske & Taylor 1984, p. 12). On the one hand, schemas condense various forms of information and certain motivational components of cognition, such as beliefs and desires. On the other hand, this form of cognitive organization also enables new information to be processed considerably more quickly (Augoustinos & Walker 1995; Bless *et al.* 2004; Fiske & Taylor 1984; Howard 1995).

Most authors assume that schemas constitute a basic principle of the organization of representations, which in turn are regarded as fundamental units of knowledge and cognition. These units can be organized not only in schematic, but also in various other ways, such as in prototypical, exemplary, and associative modes (Smith & Queller 2004). In most theories, schemas represent abstract and generalized knowledge, as opposed to contextual knowledge, which focuses on detail (Smith & Queller 2004). In a way, schemas are "mental shortcuts" reducing environmental complexity to meet the demands of everyday action. This is why schemas also have critical effects not only on the *contents* of cognition but also on cognitive *processes* such as attention, perception, and memory recall and formation.

Many theories assume that in reducing complexity, schemas first and foremost aid in *categorizing* impressions, that is allocating newly acquired information to existing schemas and assigning them a corresponding, schema-congruent meaning.

> Fundamental to this account of socio-cognitive functioning is the assumption that reliance on categorical knowledge structures is mentally easier

than the alternative of forming data-based, individuated impressions of others . . . Simply stated, categorical thinking is preferred because it is cognitively economical.

(Macrae & Bodenhausen 2001, p. 241)

In this view, schemas and their constituent categories not only reduce complexity but also provide reliable expectations of how others perceive and construct the world and make other's behavior more predictable (Bless *et al.* 2004, p. 51 f.). As such, schemas also contribute to explaining key sociological concepts such as intersubjectivity (Reich 2010) and the reciprocity of perspectives (Schütz 1953).

The activation of schemas and their underlying representations and categories takes place in a rather automatic fashion and also triggers other (potentially meaning generating) elements that are associated with the actual stimulus representation. Therefore, categorical thinking based on schemas can become problematic because categories' boundaries and the transitions between various categories are seldom clearly defined and instead tend to be fuzzy. Furthermore, many categories do overlap, so that an impression may be allocated to different categories at the same time, which may possibly contain very different contextual information (see Augoustinos & Walker 1995; Macrae & Bodenhausen 2001). Categorical-schematic processing causes events to be apprehended not only on the basis of their specific properties but also as elements of a category (e.g., gender, age, or role) and the defining attributes of this category.

In sum, schema theories provide indications of how knowledge is organized in cognitive structures and how these structures may contribute to the emotion eliciting appraisals of events. They provide a perspective that allows linking the structural, content-related aspects of appraisals with their processing characteristics and indicate how initially rule-based and deliberative appraisals can condense into automatic and schematic appraisals (Fiske 1982). Reisenzein (2001) has described this process as *appraisal schematization*. He assumes that all appraisals initially have to be computed actively or be adopted by other actors and recognized as valid and appropriate. Appraisals may subsequently be linked to other information and stocks of knowledge and then solidify into new schemas or be integrated into existing ones (Reisenzein 2001, p. 197). These schemas are then activated whenever schema-congruent information is perceived. Schematic appraisal thus always takes place when previously stored results or outcomes of appraisals are retrieved from memory and assigned to an actually occurring appraisal problem that is sufficiently similar (Reisenzein 2001).

Appraisal schematization thus offers one possible explanation for the transformation of deliberative, rule-based appraisals into case-based, associative appraisals. This transformation can also be conceived of at the neural level with reference to the two memory systems outlined in the previous section. From this point of view, schematic information processing represents a stage between controlled cortical and automatic subcortical processing of appraisals. Here, the flexibility and adaptability of information processing is due to the fact that two partially complementary systems, the cortical and hippocampal learning and memory systems, are operating in parallel. The slower but more stable cortical system contains general semantic information, while the hippocampal memory system generates only *temporary* representations of current environmental perceptions (the contextual knowledge), which can be consolidated and solidified only through regular exposure, after which it may attain the status of general knowledge (Macrae & Bodenhausen 2000, p. 94).

The importance of social cognition and the schematization of appraisals and emotions can further be illustrated by supplementing the psychological perspective with sociological views on knowledge and cognition outlined in Chapter 1. Some theories of emotion and some appraisal theories in particular claim to account for the many dimensions of sociality simply by incorporating assumptions of social cognition and schematic information processing. However, social cognition is still mostly understood as an individual or at best dyadic interindividual phenomenon.

> Some have argued that the "social" is a misnomer and that the only thing social about social cognition is that it is about social objects—people, groups, events. . . . Currently, research and theory in social cognition is driven by an overwhelming individualistic orientation which forgets that the contents of cognition originate in social life, in human interaction and communication . . . As such, societal, collective and symbolic features of human thought are often ignored and forgotten.
>
> (Augoustinos & Walker 1995, p. 3)

In this respect, cognitive sociology and the sociology of knowledge give priority to the social origins of categorical thinking and schema emergence, which are also related to ideas in phenomenological sociology, such as typification and appresentation (Berger & Luckmann 1966; Schütz & Luckmann 1973). Accordingly, representations and the ensuing cognitive structures have to be considered not only as mere "copies" of some sort of sensory input, but most of all as interpretations of them based on existing knowledge (Manstead & Fischer 2001, p. 226).

In social contexts, schematic information processing is effective and adaptive in a way illustrated above only if the underlying schemas are shared by a sufficiently large number of individuals. If I am the only individual whose actions are based on specific schemas (also known as action scripts), then (everyday) social interaction will most probably proceed in a disruptive and inefficient fashion and emotions arising from schematic processing are likely to be maladaptive and regarded as inappropriate. Surely, no two individuals will have identical stocks of knowledge that are organized in identical schemas.

However, cognitive sociology has demonstrated that knowledge is to a large extent socially constituted and that individuals have at their disposal vast amounts of comparable and institutionalized knowledge, usually acquired in corresponding social institutional settings (e.g., kindergarten, schools, the family, and dominant life-course models).

Social representations

Further support for the notion that cognitions are socially shaped and structured and originate in social and broader societal contexts comes from social-psychological work on *social representation*, which are based on Durkheim's (1915) concept of "collective representations" (Augoustinos & Walker 1995, pp. 134 ff.; Moscovici 1961). In traditional cognitive science, representations are seen as the basic elements of knowledge and cognition (Smith & Queller 2004). Actors are said to represent their environment in different representational media, for example symbolic or iconic. The purpose of representations is to "familiarise oneself with the unfamiliar" and to guide social action and behavior (Moscovici 2001, p. 20).

Moscovici (2001) argues that representations are social in origin and fulfill important social functions. In particular, they can be used to explain the links between individual actors and social units acting collectively. In order to commit one's actions to a social unit, i.e. to contribute to social institutions, public goods, and to conform to social norms, a commonly accepted system of representations is necessary. Without such a system, the emergence and coordination of collective actions and intentions would be almost impossible (Lahlou 2001). This enabling function of social representations is realized by the interactions between individuals and society through a combination of knowledge and beliefs, both of which are sustained by actors' daily practices and experiences. The notion of social representations differs from Durkheim's account of collective representations mainly by the former's emphasis on the dynamics and plasticity of representations, even though they could not persist without the possibility of at least temporary consolidation (Augoustinos & Walker 1995, p. 136 f.).

Hence, sociological social psychology offers a detailed account of one side of the process by which, according to Berger and Luckmann (1966), a common social reality emerges out of individual actions and social stocks of knowledge. Crucial to this side is the process of *internalization*, "by which the objectivated social world is retrojected into consciousness in the course of socialization" (Berger & Luckmann 1966, p. 61) and which "is the basis, first, for an understanding of one's fellowmen and, second, for the apprehension of the world as a meaningful and social reality" (p. 150). However, theories of social cognition and social representation often neglect the other side of the coin, namely processes of *externalization*, i.e. the constitution of a common social (cultural and material) reality through human activity, which is perceived to be external to the self and experienced as an objective reality (p. 150).

Externalization is composed of objectivation and institutionalization, and in social representation theory it is the former that is the primary mechanism responsible for consolidating knowledge structures. Objectivation is a process whereby the externalized products of action attain the status of objectivity (p. 60 f.), or, in other words, socially shared concepts and ideas are transformed into concrete and "objective" reality (Moscovici 2001).

In this way, the concept of social representation can serve as a robust link between psychological research relevant for emotions and the sociology of knowledge. It is based on well-established findings regarding social cognition and schematic information processing which aid in emphasizing the social construction of knowledge and mental representations. Importantly, this linkage therefore also gives insights into the social structuration of emotion based on social representations. Moreover, externalization and objectification first provide a conceptual basis for the transformation of micro-social order into objective social reality, for instance in the form of symbolic orders. Second, there is no reason to believe that externalization and objectification are limited to ideas, beliefs, and knowledge. They may equally well account for socially shared feelings and emotions, granting them the status of objective social reality.

Summary

What conclusions can we draw from the perspectives outlined in this chapter? Given that cognitions do not exist independently of actors' social embeddedness and, in turn, are a necessary component of emotions and their elicitation, then it necessarily follows that the structuration of emotion elicitation is in some way related to the social structuration of knowledge and cognition. Moreover, we can assume that emotions are subject to the principles of externalization, similar to those governing the social stocks of knowledge, and play a crucial part in institutionalization and objectivation, both as facilitators and objects of these processes. The research reviewed above suggests three possible ways in which (cognitive) emotion elicitation is subject to social structuration.

1. Through the schematization of appraisals. Like other cognitive processes, appraisals can be condensed into schemas. Initially deliberative and controlled evaluations of events consolidate into schematic structures, so that, upon perception, also complex events and social situations can rapidly trigger "schematic emotions." These emotions are internalized in much the same way as declarative and conceptual knowledge, suggesting that they are subject to comparable processes of social structuration.
2. Emotions and affective reactions can themselves become parts of schemas. Fear, for example, can be a component of a "firearms schema" or affection can be part of a "child schema." Interestingly, these appraisals are not actively computed (as in schematic emotions which are based on schematic

rather than "raw" information), but rather recalled and activated as components of existing schemas, including the corresponding emotional reaction.

3. If emotions are the consequence of appraisals comparing current situational information with general knowledge structures and motivations, then theories of social cognition and representation suggest that the appraisal-input from an immediate situation is not exclusively based on "factual" information inherent to a situation, but also on subjective knowledge associated with a situation, which is shaped to a considerable degree by pre-existing general knowledge structures and schemas.

Thus, except in the rarest cases, appraisals do not take place on the basis of "raw" perceptual data, but rather of the schemas that are activated by the perceptual process. To put it another way, the appraisal of a gun being brandished by a villain does not take place on the basis of a detailed analysis of a metallic object of a certain shape and a person whose face is masked, but rather on information associated with a "firearm in a gangster's hand" schema, which may be the result of subjective experience or culturally learned (or both). It should be noted here that the extent to which perceived information is processed by using pre-existing general knowledge structures crucially depends on existing longer-term moods (see Chapter 3).

To date, the various social influences on appraisal are just beginning to be discussed in emotion research (e.g., Mummenthaler & Sander 2012; Parkinson 2011; Parkinson & Simons 2009), even though they clearly have signifi-cant implications for the social construction and structuration of emotion. The findings presented in this chapter indicate quite clearly that there are a number of possible mechanisms, at both neural and cognitive levels, by which emotion elicitation is socially shaped and structured. Indeed, accounting for psychological and neuroscientific perspectives shows that this social shaping is considerably more far-reaching and fundamental than previously thought, a finding that to date has gone largely unrecognized in much of the sociology of emotion. By means of the mechanisms just portrayed, social order "impinges" its structural and symbolic properties onto actors' emotions, often at levels that elude conscious influence and intentional control, which is why this shaping is most effective in terms of the repercussions for social action and interaction and hence the reproduction of social order. These repercussions will be discussed in in detail the following two chapters.

3 The affective structure of social action

In the previous chapter, I have argued that a sociological explanation of emotion elicitation profits from considering both neural and cognitive levels of analysis precisely because culture and society exert marked and systematic influence on these levels. In developing this argument, I have shown that social structure and symbolic order are at least partly reflected in actors' cognitive and neural structures which in concert bring about the "structures of thought" as well as the "structures of feeling" that are characteristic for a social unit. Until now, these structures only mirror a *unidirectional* process and consequently remain at the analytical level of individual actors—from the structures of culture and society to the systematic shaping of emotion-relevant neural circuits and cognitive structures and finally to the structuration of affect and emotion. However, the main argument of this book holds that emotions function as *bi-directional* mediators between social action and social structure. To account for this second direction, this chapter will initially show how socially structured emotions in fact contribute to the emergence and reproduction of social order. The key question I will address in this chapter is this: What influence does emotion exert on social action and how is action systematically shaped by affects and emotions?

The investigation is confined initially to what Weber (1968) named "covert" action, but the basic premise is that the "internal" actions carried out in this way also have consequences for external, overt action and hence for sociality. The starting point of my analysis are two classic concepts in sociological theories of action—rationality and norms—which will be discussed by drawing on the much-debated opposition between *homo sociologicus* and *homo economicus*. However, I will not discuss these concepts and their basic assumptions in great depth (this has been done elsewhere), but will rather focus on identifying key problems and outlining possible approaches to solutions rooted in emotion research.

The following section starts with a detailed analysis of the influence of emotion on cognition, in particular on information processing, storage, and retrieval. Basically, I extend the perspective of the previous chapter to also encompass the *consequences* of emotion on cognitions and the (supposedly) rational and normative foundations of action. The relationship between rationality and emotion and the role of emotions in decision-making has become a vibrant field of research is many disciplines, and I will mainly draw on findings from psychology, cognitive science, and behavioral economics. One key reference is the now classic "somatic marker" hypothesis (Damasio 1994), which, in conjunction with other theories and empirical evidence, provides a model of how rational decision-making is fundamentally dependent on affective processes, in particular in personal and social domains. In this vein, I will also argue that many "higher" cognitive activities that are crucial for action or even constitute actions—from assessing information to making judgments—are fundamentally dependent on feelings and that this influence is by no means arbitrary but systematic along the lines of the social structuring of emotions as elaborated in the previous chapter. Finally, I will develop an ideal type of "affective action," which in contrast to Weber's (1968) ideal type rests on the assumption that action that is influenced by emotion is neither "chaotic" nor "unpredictable" but in fact contributes to everyday "orderly behavior."

Some determinants of social action

A fundamental explanatory problem in sociology is how individually motivated action and collective action interlink in the emergence and reproduction of macro-social phenomena, such as social structure, social order, and social institutions. Sociological theories of action which broadly subscribe to methodological individualism or situationalism—as outlined in Chapter 1 —advocate two basic models of action that have come to prevail over most alternative concepts, the *homo sociologicus* and the *homo economicus*. As proxies for a wider range of theories, these two "bogeymen of the social sciences" (Weise 1989) have exerted lasting influence on the sociological understanding of the connections between social order and individual action. Up to now, they have remained largely antagonistic to each other, although there has been no lack of attempts at synthesis. The notion of individual action as driven by "external," societal forces has been characteristic of sociology ever since Emile Durkheim (1938). In contrast, Weber (1968) was instrumental, more than virtually any other sociologist, in establishing the notion of action as driven by different forms of rationality. The two models are obviously based on very different assumptions about the foundations of social action: social norms and rationality. Both are assumed to have the potential to produce regularities in behavior and thus to contribute to the emergence and repro-duction of stable social structures.

Although Weber evinced (1968) four ideal types of social action—instrumental and purposeful, value rational, traditional, and affective—his own and other works successively focused on only the first two types, and Weber's classification in fact suggests that in the end there is only one principal determinant of action, namely rationality. Ever since antiquity, myths, legends, narratives, and scientific analysis have considered rationality as the most important basis of human thought and conduct that is under constant threat of being undermined by its counterparts, the passions and emotions. In this view, emotions interfere with and distort rationality, standing in the way of analytical thinking and deliberation, and should therefore be kept in check as far as possible. Not only do emotions undermine rational thought but—even worse— one surrenders almost helplessly to them, they beset one often without warning and they cannot be sought out or freely (i.e., rationally) chosen.

According to this perspective, emotions affect not only the ideal of rationality but also that of free will. This applies not only to negative emotions such as fear and anger but also to positive feelings such as love or joy. Emotions are welcome or unwelcome experiences we seek or like to avoid, to which we adapt our plans and goals, and which we enjoy and savor, or abhor and seek to suppress—they can be both reward and punishment alike. Thus, in some circumstances, emotions act as welcome advisers with regard to rational action, for example when arguments alone do not suffice or when our "gut feelings" come to our rescue in situations in which rational thought has its limitations.

Re-examining this (at least historically) widely accepted view of emotions while at the same time considering the picture of emotions that sociology has painted and, to some extent, continues to paint, the convergence between these two pictures becomes evident. In this vein, Barbalet (1998, pp. 29 ff.) emphasizes three paradigms concerning the relationship of emotion and rationality: the "conventional" approach, which is represented by Weber's view of emotion and rationality as two opposing poles; the "critical" approach, which assumes that the relationship between emotion and rationality is one of mutual complementarity; and the third, most radical approach, which sees emotion and rationality as a continuum, as James (1897) did. According to this last view, emotion and rationality are simply two separate everyday psychological concepts, two categories of thought about thinking, acting, and feeling and not a "natural category" (cf. also Barrett 2006; Griffiths 1997).

There is probably no single sentence in history that has exerted greater influence on views of human nature and the relationship between emotion and rationality than Rene Descartes's "cogito ergo sum." For Barbalet (1998) and many philosophers, this sentence is both a crystallization point for models of human thought and a starting point for subsequent generations of philosophers, psychologists, and social scientists alike whose theories eventually established the "conventional" view. This still prominent view is expressed in models of human behavior that locate virtually all responsibility for action within the individual such that action always appears to be a consequence of thought. However, there are countless forms of action that evidently do not originate in thought, but whose origins can rather be attributed to processes and physiological reactions that often occur outside awareness and cognitive control: feelings, sensations, passions, and emotions that one does not choose to have, but which simply arise without conscious involvement (Barbalet 1998; Solomon 2004).

Weber's ideal types of action, which still play a significant role in today's sociology, are characteristic of the reception given in sociology to this view of the relationship between emotion, rationality, and action.

> By "action" in this definition is meant human behaviour when and to the extent that the agent or agents see it as subjectively *meaningful*: the behaviour may be either internal or external, and may consist in the agent's doing something, omitting to do something, or having something done to him.
>
> (Weber 1991, p. 7)

This definition already suggests that the subjective meaning of action and its general interpretability, i.e. the possibility of attributing underlying intentions, motives, and beliefs, makes human action appear significantly less "irrational" than, for example, animal behavior. Consequently, rational action, and in particular purposeful instrumental action, is rational because it is based on conscious, deliberative thought and its cognitive structures and motivational states (cf. Barbalet 1998, p. 35). For Weber, emotions stand in stark contrast to instrumental rational action—they are inherently irrational and can dominate rational thinking, thereby submitting it to their very own logic. Accordingly, instrumental rational action must always be directed against emotions as spontaneous and impulsive forces that merely divert actors from their actual goals (p. 37).

According to the normative paradigm in sociology, recurring patterns of action that constitute social order can be explained by looking at the norms and conventions that constrain conduct. Parsons (1951), for example, deviated from Durkheim's strong normative stance inasmuch as he regarded norms as simply one component of sociality constituting the "action frame of reference." Parsons did not view actions as purely rational choices, but he also did not regard norms as rigid constraints—as Durkheim did—but rather as a framework guiding social action. This normative paradigm has further been developed in role theory, according to which actions are primarily guided by normative expectations, which are forces shaping the development of the social selves and identities and their corresponding behavior. Normative expectations usually are internalized in the course of socialization and thus often lose much of their constraining character (Biddle 1986; Dahrendorf 1958; Turner 1978). Whether internalized expectations or external constraints, or whether social norms, conventions, or moral norms, norm systems and categorizations provide a multitude of different forms of compliance from which to "choose"; however, actors' motives for actually adhering to norms are frequently located *outside* the concept of the norm itself—for example in contingency management, in coping with multilateral behavioral expectations, in punishment and sanctions, in individual or social functions, or in collective utility.

Furthermore, although the argument that norms and normative expecta-tions are internalized suggests that the normative paradigm ceases to act as a purely external constraint, the solutions on offer, such as creative *role-making* (as opposed to pure *role-taking*), do little to change the criticism that norma-tive behavior is "over-socialized" (Frank 1993). Also, there are a number of unresolved questions concerning the ontology of norms: What are norms? How are they cognitively represented? Why do actors ultimately comply with norms? How do norms emerge and how are they maintained and enforced in large-scale societies (see Coleman 1990; Elster 1989; Hechter & Opp 2001; Horne 2001)?

Indeed, theories of rational action have attempted to answer some of these questions. In such attempts, *homo economicus* most decidedly emerges as an actor who certainly does not live a life of dependency on *homo sociologicus* but on the contrary emphatically asserts entitlement to an independent exist-ence. According to this view, the rational actor is a counterpart to "over-socialized" conceptions of conduct and driven by goals and expected utility. One of the most insistent social science advocates of the rational actor model is James Coleman (1990). His principal criticism of the normative paradigm is that it takes as its starting point actors as *already socialized* elements of a social system. Because of this, Coleman argues, the basic question of micro–macro linkage cannot even be adequately posed within the normative paradigm. According to Coleman, the interplay between individual oppor-tunities and motives for action and the restrictions imposed by external social circumstances plays no part in the normative tradition. Thus, Coleman and other proponents of rational choice have sought to learn how norms emerge as a result of the coordination of rational action (see also Bendor & Siwstak 2001; Horne 2001; Opp 2002).

Obviously, however, the model of rational actors making decisions on the basis of subjective expected (individual or collective) utility has been criticized for various reasons from as many quarters as the normative paradigm itself (Archer & Tritter 2000; Elster 1999; Smelser 1992). Much of this criticism can be traced back to problems with empirical applicability, particularly outside of rational choice theory, or to excessive abstraction and simplification of the models, as well as to the ambivalence of the concept of rationality. Empirical studies have demonstrated over and over again that human action is characterized by several anomalies and deviates from the propositions of rational choice theory on a regular basis (Camerer & Fehr 2006; Collins 1993). Against this background, the fundamental problems of rational choice theory include, for example, inconsistencies in measuring individual and collective utility, the emergence and ontology of preferences and their transi-tivity, intertemporal effects on decision-making, the paradox of choice, and reciprocity and altruism.

Critique also concerns propositions in rational decision theory covering the establishment of and adherence to social norms. It has been emphasized, for example, that adherence to social norms turns out on occasions to be irrational, in the sense that the costs of compliance far exceed its utility, such as in the case of conventions and rules of etiquette that are obeyed even when no other actors are present (for a detailed exposition, see Collins 1993; Elster 1999; Frank 1993). Experimental studies in behavioral economics and game theory have demonstrated that standard rational choice theory is beset by coordination and cooperation problems, particularly in social dilemmas. Two well-known paradigms in behavioral game theory—the ultimatum game and the dictator game—make it clear that rational, utility-maximizing strategies are by no means the rule in experimental settings. Rather, these games suggest that in social dilemmas, norms do play a decisive role but not primarily in terms of their "rational" underpinnings, but in view of their emotional and affective components.

Apart from the general criticisms of rational choice theory, the following sections seek to show how some of the problems related to normative theory and rational choice theory in particular can be mitigated by taking account of the role of emotions in social action and "rational" decision-making. I will illustrate that affect and emotion critically interact with a variety of cognitive processes involved in rational thought. Importantly, they do so in a way that reflects their social and cultural constitution—their social structuration—and actors' social embeddedness in culture and society.

Cognition, emotion, and rationality

A good point of departure to investigate the relationship between emotion and rational action is to take a closer look at the interplay of emotions and the cognitions on which action is based. Cognitions here would include both mental contents and representations as well as information processing. From this standpoint, this section seeks to elucidate the effects and consequences of socially structured emotions for the cognitive structures and processes implicated in rational action and decision-making. As shown in the preceding paragraphs, this question has usually been answered relatively unambiguously: Affect and emotion exert a fundamental and in most cases disturbing or even disruptive influence on cognition, in particular on rational action. But what is the precise nature of these effects or repercussions? And do emotions in fact influence cognitions only by adding "noise" and thereby giving rise to "suboptimal" individual or socially dysfunctional outcomes? Or is there reason to believe that emotions also have a functional and adaptive effect for cognition and rational action?

If such effects could in fact be observed, then it seems plausible that the structural regularities of emotion elicitation also propagate into cognitive processing, altering it systematically. This would hint at reciprocal interactions between emotion and cognition with the possible consequences of strengthening the structural characteristics of both. Let us assume first that cognitive structures represent socially shared and structured knowledge, as outlined in the second chapter. Let us further assume that these cognitive structures give rise to specific kinds of affective processes and emotions—which are in turn socially structured. If we can further assume that these (and other) affects and emotions exert systematic influences on cognitions and through cognitions on everyday forms of social action, this would constitute a crucial explanatory link in view of the *reproduction* of social order. This linkage is also crucial to the social and individual *functions* of emotions. The vast majority of cognitive theories hold that emotions generally have an adaptive function realized through several intraindividual components of emotion (Levenson 1999). Likewise, several functions of emotion have been identified in social interaction, social relationships, and society at large (Frijda & Mesquita 1994; Keltner & Haidt 1999).

A wealth of empirical evidence has beyond a doubt established that emotions influence cognition:

> An individual's affective state may influence each and every step of the information processing sequence, from selective attention to information, to the encoding of information and its subsequent retrieval from memory. In addition, affective states may influence evaluative judgments and individuals' choice of heuristic or systematic processing strategies.
>
> (Clore *et al.* 1994, p. 369)

To categorize these affective influences on cognition, the extant literature differentiates between structural and processual influences. On the one hand, emotions influence *what* individuals think or *what* information they process; on the other hand, they influence *how* information is processed (Forgas 2000, p. 254).

Surprisingly, there is little agreement over the specific nature of these influences, the conditions under which certain kinds of influences occur, and the precise direction of these influences. The available evidence suggests that the disruptive effects of emotions primarily occur when emotions are intense and involve high levels of physiological arousal. Such emotions are said to make us "take leave of our senses," to leave us "blinded by love" or "sick with jealousy." However, besides these strong and usually obvious influences, there are more subtle emotions, moods, and affects that are often not at the center of conscious awareness and might make us do the right thing "intuitively," so that in certain situations we act "correctly" but without ultimately being able to give a reason for our actions. Similarly, we may "take an instant dislike" to someone or feel ill at ease in certain places or situations that "objectively" give us no cause to do so. And we sometimes take products off the shelves in supermarkets for which we may have no use at all, or we "decide" to take the more expensive of two alternatives, which, from a rational choice perspective, undoubtedly provide the same utility (Glimcher 2003).

At a neural level, these phenomena can be explained by automatic and non-conscious processes in the basic affect system and their interactions with other higher cognitive processes involved in rational thought, decision-making, and action—as is done by the somatic marker hypothesis (Damasio 1994). At the cognitive, representational level, psychological emotion research has developed several theories that aim at explaining the precise nature of the influence of emotions on cognition. I will discuss both the neural and cognitive-representational perspectives in the subsequent sections.

Affective information processing

In most existing models, the effects of emotion on cognitions are located primarily in memory, judgment and decision-making, and information processing. The majority of existing theories do not discuss the effects of discrete emotions, such as joy, anger, or fear, but rather the effects of moods and feelings according to their positive or negative valence. Here I will briefly review three well-known and related paradigms: mood congruency, mood as information, and affect infusion.

Mood congruency

Looking at the role of emotion in storing and retrieving information from memory, Clore and colleagues (1994) emphasize the importance of attentional deployment and the valence of occurring moods and emotions. They assume that sufficiently strong emotions (regardless of their valence) make demands on sensory and phenomenal attention, thereby reducing the available resources for other stimuli that are less salient and emotionally relevant. In a comprehensive review of existing research, they conclude that negative affect reduces the elaboration and organization of information at the encoding stage and also adversely affect retrieval of that same information (Clore *et al.* 1994, p. 372). This seems to be particularly the case when the information that is encoded is in itself affectively neutral (e.g., when studying for an exam). However, when the encoded information bears strong negative or positive connotations, both strengthening and weakening effects on storage and retrieval have been observed (Clore *et al.* 1994, p. 372).

These occasionally contradictory effects are frequently explained by reference to the concepts of "mood congruence" and "state-dependent recall." The latter refers to findings indicating that the matching of affective states at encoding and retrieval time facilitates memory recall independently of the valence of the material that is encoded and recalled. In contrast, the former holds that the matching of the affective state at recall with the valence of the material that is remembered facilitates retrieval (Clore *et al.* 1994, p. 376). Research on state-dependent retrieval has shown that retrieval is significantly improved when the situational contexts at the time of encoding and remembering are congruent, and emotions are of particular importance as indicators of a particular context. State-dependent recall has been shown to come into play especially when:

- a situation at the time of recall provides only few clues to help identify a specific context;

- stored information has a personal relevance; and
- affective states at the time of both storage and recall are relatively intense.

(Berkowitz 2000, pp. 67–95; Bless *et al.* 2004, pp. 179–183; Clore *et al.* 1994, p. 375)

Mood congruence, on the other hand, has been demonstrated primarily for autobiographical memory and has otherwise been hard to validate. A number of studies have shown that people in a happy state remember positively connoted information considerably better than negative ones (and vice versa). However, as with state-dependent recall, differentiating between the valence of stored information and the current affective state is problematic: It is plausible that the occurrence of "positive" events triggers equally positive affective states at encoding time, such that the valence of an event is *de facto* congruent with the affective state at encoding, and it becomes virtually impossible to distinguish between the two aspects. Given the inconsistent evidence, it is safe to assume that mood congruence effects are most likely to occur when criteria for mood congruence and for state-dependent recall are given (Clore *et al.* 1994, p. 376).

In explaining both mood congruency and state-dependent recall, research frequently refers to associative network models of memory and information processing (Anderson & Bower 1973). In such models, emotions can be represented as nodes that are connected to other nodes representing events, physiological reactions, appraisals, or other contextual cues associated with an emotion. Information that is encoded under a particular affective valence becomes associated with other network nodes that have already been active at encoding. As a result, activation of an emotion node in the network also tends to activate associated network parts and activity in one part of the network can rapidly spread between associated nodes (*spreading activation*) (Clore *et al.* 1994, p. 373). Using associative networks, state-dependent recall is explained by assuming that information stored under a certain affective valence becomes associated with a corresponding emotion node. If this node is also active during retrieval, this facilitates recall of associated material.

Critics argue that associative networks merely translate empirical phenomena into the symbolic language of networks and that this translation is close to arbitrary (Schmidt-Atzert 1996, p. 205). Also, Clore and colleagues (1994) express doubts concerning the representational media on which associative networks are presumed to rely. Bower (1981) identified six basic emotions that he assumes to be biologically hardwired and represented at neural levels as nodes in an associative network. These nodes can activate both physiological reactions (represented as motor codes, for example) and other network nodes that point, for example, to propositional content. However, it seems problematic to assume that emotions *themselves* are represented as nodes in associated networks. Rather, it is plausible that the *socially and culturally constituted concepts* of discrete emotions are represented as nodes in such a network. From this point of view, it is not emotions *as such* that influence information encoding and retrieval, but rather the activation of corresponding *concepts* of emotions (Clore *et al.* 1994, p. 374). Looking at the previous chapters, there is much to suggest that the elicitation of early affective responses, which is almost exclusively based on non-declarative memory, happens on a dimensional valence and arousal continuum rather than in terms of the basic emotions postulated by Bower (1981). This view would hold that initial and basic affective responses subsequently activate certain semantic emotion concepts, which then trigger the network effects described above.

This view of the representation of emotions in associative networks lends support to the arguments advanced in the previous chapters in that it avoids the reductive determinism of basic emotions and instead allows for a degree of social and cultural variability in the representation of emotion. It also provides theoretical support for certain assumptions made by multi-level appraisal theories, some of which make explicit reference to associative networks (Smith & Kirby 2000). Another view is that emotions and affective states can serve as (propositional or non-propositional) input to—in particular automatic—appraisals and simultaneously function as context-discriminating factors that activate associated material in a network that serves as additional appraisal input (Clore & Ortony 2000; Reisenzein 2001). This perspective suggests an understanding of emotion nodes as pre-existing schematic appraisals in the form of propositional codes. In this regard, Clore and Ortony (2000) note that "activated material, be it emotional or not, can be structurally complex and highly organized, so that accessing any part of a structurally complex representation (or schema) may have extensive implications" (p. 35). In summary, it may be said that encoding and retrieval of information to a large extent depend on:

- current moods and affective states;
- the valence of the encoded information; and
- the representational structure and media of memory.

For the overall argument advanced in this this work, it is crucial that affects and emotions systematically influence action and behavior by making available and possibly bringing to conscious awareness information—i.e. autobiographical experiences, implicit and explicit memories, declarative knowledge, beliefs—that is associated with and potentially relevant to the current situation. Importantly, emotions selectively provide such information to an ongoing situation (e.g., action planning, judgments, decision-making, habitual behavior) that is affectively congruent with the current emotional state. Thus, rational thought, judgment, decision-making, and other forms of social action operate with information that tends to match the valence of prevailing emotions and hence is highly selective and dependent on past experiences. Given that the emotions exerting this influence are elicited in socially structured ways, they constitute a crucial link between structural and symbolic orders and social action.

Mood as information

While mood congruency primarily contributes to explaining *what* actors think and what information is recalled and processed, other models focus on *how* information is processed depending on current affective states. Given the interactions of mood and memory, it seems reasonable that judgment and decision-making are not immune to affective influences. Mood congruence and state-dependent recall influence which information is made available to serve as input for the decision-making process. This indeed reflects our own everyday experience, namely that judgments and decisions take place in accordance with prevailing emotions in a given situation. Thus a romantic relationship, for example, is assessed more positively if one is in a positive mood at the time of evaluation, while it is difficult if one is in a bad mood to be amused by jokes and stories that under different circumstances would cause joy (Forgas 1995). Similar effects have been observed with regard to the activation of stereotypes and the adoption of attitudes (Bless *et al.* 2004, p. 188), the evaluation of consumer goods (Yeung & Wyer 2004), and the perceived quality of life (Clore *et al.* 1994).

Concurring assessments and evaluations of this kind are frequently adduced as evidence in support of the mood congruence model; however, they are difficult to substantiate because of the conceptual inconsistencies highlighted above. Here, the "mood as information" (MAI) paradigm offers alternative explanations. This model, originally developed by Schwarz and Clore (1988), is based on the assumption that, in judgment and decision-making, actors frequently resort to their own *self-perception* as a crucial source of information and, in doing so, give particular weight to their current affective state.

The MAI model holds that *affective feelings* (for instance, "gut feelings") serve as relevant and highly salient information in decision-making in their own right. The model's key proposition is consistent with the assumptions of mood congruency and predicts that evaluative judgments will tend to be more positive if actors are in a positive mood and more negative when in a negative mood (Schwarz & Clore 1988). More recent studies have also accounted for discrete emotions and specific appraisal tendencies in addition to affective valence as a decisive factor (Clore & Storbeck 2006; Lerner & Keltner 2000; Storbeck & Clore 2007). The paradigm is based on the principles of heuristic information processing in decision-making. This kind of processing usually accounts only for a limited fraction of the total sum of available information in decision-making and instead relyies more on "rules of thumb" than on detailed analysis of information (Gigerenzer 2007; Gigerenzer & Gaissmaier 2011). Heuristic information processing generally occurs in relatively "unproblematic," everyday situations (and therefore seems to be of particular importance in view of the reproduction of social order) as well as in situations in which the resources and information required for a comprehensive analysis are unavailable (see Berkowitz 2000; Bless *et al.* 2004; Schwarz & Clore 1988). Most of the automatic and schematic modes of appraisal processing discussed in the previous chapter are also based on heuristic information processing.

Crucial to MAI influences is the attribution of a subjective feeling to an intentional object. While this attribution can be closely linked to the circumstances that gave rise to the feeling in the first place (or some condition for its existence in the narrower sense), moods are often characterized by the absence of an intentional object. This is why certain affective feelings of longer duration and less intensity than emotions proper, such as moods and sentiments, are often attributed to objects that are not the cause of the affective state (misattribution). Nevertheless, these feelings are accounted for in evaluative judgments regardless of the "correct" or "incorrect" attribution (cf. Clore *et al.* 1994, p. 381; Schwarz & Clore 1988). Studies have demonstrated that MAI effects do not occur when alternative causes of an affective feeling can be identified or when no associations can be established between mood and the situation or event that is to be evaluated (Bless *et al.* 2004, p. 184; Clore *et al.* 1994, p. 381 f.).

In summary, the influence of feelings on information processing in judgment and decision-making can be explained in two ways. First, moods and feelings that are (mis-)attributed to an event that is evaluated are of informational value in themselves. I am feeling good about something and therefore I tend to evaluate it positively—even if this something is not the cause of the feeling. Second, feelings facilitate the recall of and access to stored information of similar valence, which in turn contributes to a valence-consistent evaluation of the situation. If I am in a good mood, I tend to recall information that is positively connoted as well, and this information facilitates the positive evaluative judgment.

Affect infusion

A third approach to the influence of emotion on cognition aims at specifying what kinds of effects will occur for what kinds of affective feelings. The "affect infusion model" (AIM; Forgas 1995, 2000) makes assumptions on both the contents of cognition (*what* actors think) and the mode of information processing (*how* actors think). In essence, the model postulates that affective influences on evaluative judgments and decision-making mainly occur in complex and exceptional circumstances. Forgas (2000, p. 255) distinguishes four basic classes of information processing: direct access, motivated, heuristic, and substantive processing. The first two modes leave virtually no scope for affective influences. Direct access processing falls back on rigid stimulus–response patterns, in particular in familiar and frequently recurring situations characterized by habitual behavior. Here, the close links between incoming information and the activation of adequate reactions makes it virtually impossible for affective feelings to exert much influence. Motivated inform-ation processing is characterized by a particularly strong focus on specific goals or objects and hence is highly selective, leaving little room for affective influences. However, emotions are a frequent *trigger* of this kind of processing. In heuristic processing, feelings are particularly influential when misattributed to an event or object (similar to the MAI paradigm).

Evidence suggests that the substantive, detailed, and resource-intensive processing of information (which is comparable to rule-based, non-schematic processing postulated by appraisal theories) is particularly susceptible to the influence of moods and feelings (Forgas 1995, 2000). This becomes evident, for example, in complex social situations that demand coordination and cooperation, in instances of strategic behavior, or more generally in problematic and unexpected situations (Forgas 2006). Specifically, the AIM proposes that positive feelings foster top-down, schematic, and heuristic processing, while negative affects promote detailed bottom-up processing.

Feelings and general knowledge structures

An alternative view of the role of affective feelings in information processing, in particular problem-solving, shares the assumption that positive emotions lead to the use of relatively superficial and simple heuristics (Schwarz & Bless 1991). However, in situations requiring more analytical and in-depth processing, positive emotions may lead to poor problem-solving outcomes. In situations requiring innovative and creative solutions, they tend to lead to better results. Schwarz and Bless (1991, p. 59) depart from the basic assumptions of the MAI paradigm and extend it regarding the role of affective valence. According to their view, positive affect indicates relatively safe and unproblematic situations, while negative affect signals problematic and uncertain situations (Schwarz & Bless 1991, p. 59).

Actors in a positive mood therefore tend to construe situations as unproblematic and prefer simple and resource-saving heuristics over detailed analytical processing.

> Specifically, we suggest that *positive moods* are likely to elicit a processing strategy that relies heavily on the use of simple heuristics, and that is characterized by a lack of logical consistency and little attention to detail ... In contrast, we suggest that negative moods are likely to elicit an analytical mode of information processing that is characterized by considerable attention to detail ... and a high logical consistency, although probably associated with a lack of creativity.
>
> (Schwarz & Bless 1991, p. 56)

Bless (2000, 2001) develops an interpretation of this argument that is of particular interest from a sociological perspective. He assumes that positive moods as indicators of unproblematic situations lead to the activation of *general knowledge structures*. These stocks of general knowledge are represented in schemas and prompt well-established and validated (standardized, habitual) courses of action that have proven to be adequate and successful in comparable situations (top-down processing). On the other hand, negative moods indicating problematic situations tend to reduce reliance on general knowledge structures and existing situational representations, schemas, and scripts, since these are probably less appropriate in unusual and problematic situations.

Thus, negative moods motivate actors to pay particular attention to and to process more intensively information that is inherent in the situation they are *currently* experiencing (bottom-up processing). Positive affect tends to encourage actors to fall back on general knowledge structures while neglecting information inherent to the current situation (Bless *et al.* 1996; Bless 2000, pp. 204 ff.). Fiedler and Bless (2000) have speculated on the more far-reaching implications of these two basic processing strategies for the cognitive system. They argue that the activation of and increased attention to general knowledge structures initiated by positive moods foster the modulation and transformation of newly perceived information through these general knowledge structures. Negative moods and a focus on situationally available information (emanating from the immediate environment or working memory) give rise to a "conserving" processing strategy in which attention is paid to detail and little scope is left for the influence of existing stocks of knowledge to modulate incoming information.

This social cognitive process is basically constituted by two components: On the one hand, the simple and unaltered perception of an event on which judgments or decisions are based (*conservation*). "Conservation means keeping a record of whatever data input is perceived in the external world or retrieved from internal memory" (Fiedler & Bless 2000, p. 147). On the other hand, social cognition also involves the transformation of newly acquired information by pre-existing cognitive structures (*active generation*). Fiedler and Bless's (2000) core argument is that "[p]ositive emotional states facilitate active generation, whereas negative emotional states support the conservation of input data" (p. 147).

This view is closely related to Piaget's (1954) dialectic relationship between the two adaptive processes "assimilation" and "accommodation." The active use of existing knowledge structures is similar to the assimilation strategy, a top-down process in which what is perceived is adapted to fit existing models. Assimilation thus reflects confidence and trust in existing knowledge structures (e.g., norms, values, or behavioral strategies) (Fiedler & Bless 2000, p. 145). The conservation strategy, similar to accommodation, relies more on newly acquired information and in fact "conserves" it to restructure and update pre-existing knowledge.

> People in negative states should be more cautious, suspicious and careful than elated people in judging . . . The emphasis they place on conservation should lead to a detailed observation process, longer hesitation before making a decision, a final decision that is highly predictable from the stimulus input, and a rather systematic diagnosis. Conversely, decisions made in a positive mood should be faster, more detached from stimulus-constraints, and more prone to knowledge-based inferences.
>
> (Fiedler & Bless 2000, p. 148)

Critically for a sociological analysis, the general knowledge structures in question can be assumed to reflect the social structures within which they arise or—even more—constitute the social structures of thought. Thus, in the light of the theories expounded in Chapter 1, we can summarize that social structures and their cognitive representations play a decisive role in determining the influence of emotion on cognition.

Implications for emotion elicitation

Applying these theoretical arguments and empirical findings in turn to the role of cognition in the *elicitation* of emotion, a picture of extensive recursive interdependencies between cognition and emotion emerges. By definition, appraisal processes, whether controlled or automatic, are cognitive processes that relate to an actor's overall well-being. Accordingly, appraisals—like other cognitive processes—are susceptible to the influence of existing moods and feelings. These feelings contribute to the (mood congruent) selection of information that is used as appraisal input, thus decisively influencing the elicitation of immediate emotions proper. In cases of positive feelings, perceived information is adapted to pre-existing knowledge structures (assimilation), and appraisals are therefore based at least partly on modulated and transformed (schematic) input and not on "raw" perceptual data. Moreover, mood congruence and state-dependent recall provide ongoing appraisal processes with information that is both valence-specific and highly selective, thereby ensuring that the general knowledge structures that are activated correspond—at least in terms of valence—with the prevailing mood (Fiedler & Bless 2000, p. 149). Negative mood, on the other hand, contributes to controlled and active appraisals with conservation and in-depth processing of situational information. Therefore, it is not only the various appraisal dimensions that determine emotion elicitation. Through their influence on the cognitive system, longer-term affective feelings such as moods and sentiments also play a part in the elicitation of episodic and short-term emotions.

In summary, the influences of moods and feelings on information processing not only fundamentally affect the principles of rational choice and all forms of action that in some way involve cognitions, but also serve to enhance our understanding of the recursive links between emotions and social structures. They clearly show how the social shaping of emotions propagates into cognition and information processing, which constitute the basis of most forms of social action. They also provide a solid foundation for pursuing the question of how socially structured emotions—through actions—in turn reproduce symbolic and structural social order. The following section exemplifies this influence regarding a specific form of action, namely decision-making, and a well-established concept that accounts for the autobiographical plasticity and experience-dependency of these influences predominantly on the neural level—the somatic marker hypothesis.

Somatic markers

The somatic marker hypothesis (Damasio 1994) draws on findings related to the neural basis of affective processing and the malleability and responsiveness of this basis to subjective experience and social learning. Basically, it states that decision-making is essentially dependent on and guided by certain kinds of affective feelings (Bechara 2004, p. 30; Dunn *et al.* 2006). Damasio (1994) originally developed the hypothesis based on two prominent cases in medical history that suffered similar damage to certain cortical areas of the brain (the frontal lobes) (Damasio 1994; Macmillan 2000). Although both showed substantial brain damage, basic capabilities such as attentional control, perception, learning, memory, language, logical reasoning, and analytical abilities remained intact. However, these patients showed drastic changes in previously normal social behavior. Shortly after the incidents that caused the brain damage, they seemed to have lost the capability to act in accordance with social norms and conventions and to show a sense of responsibility. Moreover, they were impaired in making rational decisions concerning their personal and social well-being and at the same time showed a general lack of interest in the personal and social consequences of these decisions (Damasio 1994, pp. 19 ff.).

Another remarkable change was directly related to emotional behavior. The patients showed no visible emotional reactions to their personal situations and reacted without any emotional involvement to personal setbacks, losses, or decisions with dramatically negative consequences (Damasio 1994, p. 45). Interestingly, the brain damage had neither led to loss of declarative knowledge about social norms and conventions nor to the loss of access to this knowledge or the basic capacities for rational thought, attentional control, or the processing of this information in working memory. Also, the deficiency obviously did not manifest until the later phases of a decision-making process, at precisely the point at which decisions are converted into actions.

These and related cases of brain-injury (e.g., Kringelbach & Rolls 2004, p. 351 f.) show that "reduced" emotionality—particularly in terms of complex and social emotions—is associated with the inability to make rational decisions in certain situations and to implement corresponding actions. In view of rational decision-making concerning everyday situations and personal or social future outcomes (immediate or long-term), the functioning of certain cortical brain areas which are central to decision-making seems to depend on affective processes, in particular those in higher cortical areas of the brain (Bechara 2004; Bechara *et al.* 2000; Hornak *et al.* 2003).

The somatic marker hypothesis seeks to explain these empirically documented cases. It assumes that in decision-making and before any conscious cognitive decision following rational deliberation is made, the possible consequences of a particular decision option are automatically paired with a specific pattern of physiological arousal that, under certain circumstances, is perceived as a positive or negative feeling (akin to the proverbial "gut feeling"). This autonomous physiological reaction (or the perception thereof as a subjective feeling) highlights certain consequences of a possible decision option and thus reduces the possible alternatives available to a subsequent deductive decision-making process.

According to this hypothesis, somatic markers facilitate more rapid and efficient decision-making, while their absence reduces the efficiency of decision-making or even, under certain circumstances, makes it impossible:

> somatic markers are a special instance of feelings generated from secondary emotions. Those emotions and feelings have been connected, by learning, to predicted future outcomes of certain scenarios. When a negative somatic marker is juxtaposed to a particular future outcome the combination functions as an alarm bell. When a positive somatic marker is juxtaposed instead, it becomes a beacon of incentive. This is the essence of the somatic-marker hypothesis.
>
> (Damasio 1994, p. 174)

This definition makes clear that somatic markers are not a substitute for rational or deliberative decision-making. Rather, they act as a weighting mechanism for different decision options, making available additional information that can be used to evaluate the entire spectrum of possible options and consequences of decisions, which in social and personal situations is particularly wide-ranging and often underdetermined (Hinson *et al.* 2002). In principle, this function can be regarded as the neural and often non-conscious equivalent to the MAI paradigm.

Functionally speaking, somatic markers are based on the neurophysiological bases of what Damasio calls "primary emotions," which combine certain characteristics of stimuli with certain physiological reactions, and on "secondary emotions," which result from the combination of socially learned stimulus categories with primary emotions. In distinguishing primary from secondary emotions, Damasio is not mainly concerned with issues of the "basic emotions" or "core affect" debate, but rather with the nature of the representations on which emotions are based. Primary emotions are based on innate representations of certain stimulus properties, for example a certain movement, size, smell, or sound, and not on semantic or conceptual representations. Secondary emotions, on the other hand, are based on acquired and socially learned representations, which are combined with certain components of one or more primary emotions. From this perspective, therefore, somatic markers are internalized during socialization and regular interactions in stable social environments through the combination of *certain categories of (social) stimuli with certain categories of somatic (affective) states* (see Damasio 1994, p. 177).

Thus, the tight coupling of basic affective processes based on innate representations and internal preferences with the social environment is a decisive factor in the operation and experience-driven acquisition of somatic markers. According to the somatic marker account, the basic affect system is equally implicated in processing complex social situations as in supporting bodily homeostasis. Regarding the elicitation of emotions, this account is in many ways comparable to Rolls's reward and punishment system (Rolls 1999) and Cacioppo's appetitive-aversive system (Cacioppo *et al.* 2004). In particular, Rolls's concepts of primary and secondary reinforcers in many ways mirror primary and secondary emotions.

In decision-making, the social environment provides various options for and establishes contingencies of social action and the immediate and long-term reward and punishment values of decision options. Most importantly from a sociological perspective, the formation of somatic markers not only depends on actual "first-hand" subjective experience. It is also facilitated by individuals who serve as "intermediaries" between specific stimulus–reinforcement contingencies in that they represent and enact *institutionalized* and *socially conventional* forms of behavior:

> Early in development, punishment and reward are delivered not only by the entities themselves, but by parents and other elders and peers, who usually embody the social conventions and ethics of the culture to which the organism belongs. The interaction between an internal preference system and sets of external circumstances extends the repertory of stimuli that will become automatically marked.
>
> (Damasio 1994, p. 179)

The somatic marker hypothesis has been tested extensively using an experimental paradigm known as the "gambling task" that examines decision-making under uncertainty (Bechara *et al.* 2000). The results mostly support the hypothesis and also give insights into the neural correlates of the processes involved. The prefrontal cortex is particularly implicated in the functioning and development of somatic markers. Prefrontal areas receive information of virtually every sensory modality, including the somatosensory. Furthermore, certain areas of the prefrontal cortex receive information via bio-regulatory processes, such as neurotransmitter release, amygdala or hypothalamus activity (Damasio 1994, p. 181). Crucially, the prefrontal cortex not only processes information from virtually any physiological and cognitive sub-system, but also establishes *links* between representations of situations stored in declarative and implicit memory and the affect-specific somatic states associated with these situations. The prefrontal cortex thus acts as a "convergence zone," in which information from different sources, modalities, and points in time are linked as somatic markers.

The prefrontal cortex's role in the development and functioning of somatic markers establishes a view on decision-making that explicitly accounts for past experiences and socialization conditions and clarifies, mostly at a neural but also at a conceptual level:

• how categories of (social) situations are paired with categories of physiological reactions that are characteristic of affective states and emotions;

- what consequences these associations have yielded in an individual's past; and
- what consequences can be expected for future decision outcomes.

Importantly, Damasio and other proponents of the hypothesis state that the role of physiological reactions in the somatic marker model is frequently "simulated" by specialized *cognitive representations* of these reactions, which Damasio (1994) terms "*as-if* body loops." These as-if body loops develop over the course of ontogenesis, can be recalled in situations that match certain encoding criteria, and serve as resource-saving substitutes for actual physiological reactions (Bechara 2004; Bechara *et al.* 2000; Damasio 1994). Taken together, the somatic marker hypothesis provides a valuable explanation for the inability of people with certain deficits in emotional behavior to make "rational" decisions in domains of personal and social relevance, despite their maintaining largely normal cognitive capacities and executive functions. It provides a link between rational decision-making under specific circumstances and the emotional reactions to decision options and their possible consequences. The hypothesis does not, however, postulate, as critics have repeatedly maintained (Elster 1999; Panksepp 2003b), that emotions constitute a necessarily and fundamental condition for rational decision-making in all cases. On the contrary, Damasio explicitly restricts the validity of his hypothesis to certain domains and emphasizes the *supporting* role of emotions in rational decision-making.

Consideration of somatic markers and equivalent mechanisms may circumvent well-known problems of rational-choice theory, not least because it is essentially based on criticism of these theories. Obviously, in a purely economic context, decisions based on subjective expected utility and rational calculus may still seem viable. But even under the restrictive conditions of economic models of choice, with a severely limited number of possible alternative decision options, problems may arise that lead to suboptimal solutions or even to unresolvable situations, as in cases of imperfect information, uncertainty, or risk. These factors matter all the more when decisions are made in personal or social domains in which the waters are further muddied by the difficulty of assessing and comparing expected utility, possible options, time constraints, external effects or unintended consequences.

In these situations, somatic markers act as shortcuts in the decision-making process. Crucially, these shortcuts do not function arbitrarily, but are closely linked to previous experiences, learning, and socialization. This becomes evident when looking at the arguments outlined in the preceding chapter: According to this view, the operation of somatic markers is influenced by the neural plasticity of the affect system, the cognitive consequences of moods and feelings, and the overall social structuration of emotion. Damasio's (1994) perspective on emotion and decision-making is of particular interest for sociologists because it is *not* based on innate or hardwired affective reactions (primary reinforcers), but rather on secondary emotions that are acquired and internalized during socialization. It illustrates the reciprocal links between biological predispositions and social influences without lapsing into any nature–nurture dualism.

As the result of subjective experience and learning in stable social environments, somatic markers combine certain classes of social stimuli with certain classes of basic affective reactions and are a crucial element in the social structuration of emotion. If, then, socially structured emotions decisively influence (rational) social action, we can reasonably assume that the structures of feeling are also reflected in one or another way in structures of social action. Thus, at this juncture, structures of feeling bring about structures of social action.

The sociology of affective action

The well-established interplay of cognition and emotion in decision-making provides a basis for the specification and extension of certain traditional social science accounts of human action. Closer examination of the concept of rationality that is widely used in rational choice theory and rationality-based approaches to social norms suggests three optimality conditions be met that qualify action as instrumental-rational (see Elster 1999, 2004): First, decisions and actions must be optimal given certain beliefs and desires; second, they must be optimal in the light of the available information; and, third, the resources invested in the acquisition of information must be optimal given desires and beliefs about information costs and benefits. Given these conditions for rational action, scholars have identified a number of caveats (Elster 1999; Frank 1988; Loewenstein & Lerner 2003):

1. breaks in the chain of the three optimality conditions;
2. problems arising from the uncertainty and indeterminacy of situations; and
3. issues related to the acquisition, quality, and consistency of information.

In view of these potential pitfalls, some have argued that rationality and emotion are not always as mutually exclusive and disruptive as common wisdom holds. For example, individuals may evaluate emotions themselves by means of certain optimality conditions, i.e., assess them against rational criteria in the same way as actions. Instead of the criteria used to evaluate rational actions, we can also assess emotions using criteria usually called upon to evaluate the rationality of beliefs (see Elster 1996, p. 1392; de Sousa 1990). Emotions can also be seen as critical elements of one's preference structure and the seeking or avoiding of emotions leads to increases or decreases in subjective utility ("I'm really going to enjoy myself this evening!"). Moreover, anticipated emotional outcomes of a certain decision option do influence rational decision-making ("I'll only get more angry if I do that!"). Emotions also serve as (rationally considered) information in cases of indeterminacy or incomparability of outcomes. Finally, they can influence the formation of beliefs that are integral to rational decision-making—for better or worse (see Elster 1996, p. 1391 f.).

Much of the criticism regarding rational choice theory focuses on the first problem domain, i.e. the undermining of rationality through breaks in the chain of optimality conditions. In fact, these breaks frequently result from biased or distorted cognitive information processing that does not meet the above-mentioned criteria. Consistent with the conventional view of emotions as "sand in the machinery of [rational] action" (Elster 1999, p. 284), emotions can no doubt be a major cause of these biases and distortions. For example, they may undermine the rational formation of beliefs, the credibility of information (in the case of love and jealousy, for example), or the potential consequences of actions (such as in the case of pride and arrogance). Similarly, emotions seemingly disrupt rational belief-formation when judgments are made against one's better knowledge (as in cases of grief or mourning) or when craving for intrinsically aversive emotions (e.g., guilt, penance). However, these distortions are often confined to relatively intense emotions (Elster 2004, p. 32 f.).

Looking at second- and third-category problems, emotions are more often than not *conducive* rather than disruptive to rationality. In view of uncertainty and indeterminacy, the discussion of the somatic marker hypothesis has already indicated the supportive role of emotions. In terms of the third category (the optimal acquisition of information), the anticipated emotional outcomes of decisions are a crucial factor (Kuhnen & Knutson 2005; Loewenstein *et al.* 2001; Winkielman *et al.* 2007). The role of such *expected emotions* has also been discussed in sociological theory. Turner (1999a) and Collins (2004), for example, assume that a fundamental motive of social action is the pursuit of positive and the avoidance of negative emotions. Both argue that individuals consistently seek interactions based on their expected emotional outcome and the seeking of "emotional gratification" contributes to the emergence of interaction ritual chains and micro-social order (Collins 2004). Given that the principles of gratification through positive emotions are a human universal and thus equally distributed across society, interaction rituals evolve around expectations and possibilities for gratification and thereby also make a contribution to the establishment of larger-scale macro-social structures (e.g., Collins 2004; Hammond 1990; Lawler 2001; Lawler & Thye 1999).

It thus seems as if classical principles of rational choice theory were easily reconcilable with the view that expected emotions systematically influence rational behavior. Emotion-based critique of rational choice theory therefore is often countered by arguments suggesting that emotions can be perfectly integrated into the rational choice paradigm as part of an actor's preference order and hence a component of subjective utility, true to the motto "if I decide to take the green shirt, I'll be happier than with the red one."

However, this position has been criticized as "consequentialist" in that it is concerned solely with the absolute outcome *after* a decision is made (Loewenstein & Lerner 2003). Yet expected emotions can influence decision-making in many more ways, as is particularly evident in two domains: decisions under uncertainty and intertemporal choice. The view that decisions are made solely on the basis of utility expected after a decision is indeed somewhat unrealistic. Instead, decision-making is also influenced by emotional reactions towards *relative changes* in one's current situation (Loewenstein & Lerner 2003, p. 622). Other important influences are emotions associated with counterfactual thinking, e.g. from consideration of something that is not, but could have been the case, leading to either relief or regret. Both examples clearly show how expected emotions become part of the decision-making process in a non-consequentialist way.

Looking at intertemporal choice, the influence of expected emotions cannot always be clearly distinguished from that of immediate emotions, as in the case of somatic markers. A critical feature of intertemporal choice is that some future utility is "discounted" in a way similar to discounting in finance, i.e. future utility is discounted at a rate depending on when it is expected to occur (Loewenstein & Lerner 2003, p. 625). Generally, research has shown that actors prefer immediate to delayed utility. According to standard models of intertemporal choice, an increase in utility delayed for two months is discounted in the same way as a delayed utility occurring not after 21, but after 23 months. Contrary to these assumptions, however, evidence suggests that delayed gratification is not discounted in this way. Delays in utility are perceived as more severe when located in the near rather than the distant future. This is captured by the concept of "hyperbolic discounting," i.e. discounting not based on a constant but rather on a decreasing factor (Loewenstein & Lerner 2003, p. 625). Hyperbolic discounting may thus foster spontaneous and impulsive actions, although in standard models it still remains unclear what sorts of expected utility gives rise to such spontaneous actions. For Loewenstein and Lerner (2003), both expected and immediate emotions are critical in this regard, since impulsive actions frequently occur in intense motivational and emotional states.

Emotions further influence intertemporal choice by affecting preferences and tastes. Normative decision theory postulates that decisions are ultimately based on the tastes and preferences for specific outcomes of a decision. It has been shown, however, that tastes vary over time and that actors consequently make systematic errors in predicting their own preferences at the time when the consequences of a decision manifest. The primary explanation given for such prediction errors are inappropriate evaluations of one's ability to adapt and overestimations of an event's hedonic significance (Loewenstein & Lerner 2003, p. 626).

This brief review of the influence of expected emotions on decision-making indicates that emotion and rationality are not always diametrically opposed to each other. Rather, research suggests, first, systematic links between expected emotions as an element of subjective expected utility and, second, that expected emotions play an important part in the actual decision-making process primarily because they:

- prompt assessment of decision-options based on the immediate situational context rather than solely on anticipated future outcomes;
- play a role in counterfactual thinking, thus accounting not only for expected utility, but also for biases in weighing and assessing decision options; and
- contribute to hyperbolic discounting of future utility.

Given these influences, it is paramount to emphasize that whether some expected emotion is indeed desired or eschewed, whether it is deemed as something good or valuable or rather something that should be avoided by all means, is critically dependent on the cultural meanings of emotions that are shared within a social unit. It would be too simple to assume that people simply seek out "positive" emotions such as joy or pride and avoid "negative" ones such as anger and fear. In fact, research is just beginning to establish "what people want to feel and why" (Tamir 2009), thereby highlighting cultural values as a most important determining factor. Think of the historical variability of valued and de-valued emotions, with melancholia and coolness as examples at hand (Frevert 2011; Radden 2000; Stearns 1994).

Aside from these linkages between expected emotions and the conditions for rational action, the influence of immediate emotions on action seems to be even more substantial, as indicated by the studies on cognition–emotion interactions reviewed in the preceding sections. In view of rational social action and decision-making, the influence of immediate emotions can be divided into direct and indirect effects (Loewenstein & Lerner 2003). Indirect effects include emotion-induced changes in information processing, judgments, memory formation and recall, as already illustrated. The direct effects of low- and moderate-intensity emotions and moods have been reviewed extensively in relation to the "mood as information" paradigm. In accordance with the conventional view, we have also noted that the direct effects of intense emotions may be detrimental to rational decision-making. However, this does not mean that such effects have no influence on the emergence and reproduction of social order. The sociology of emotion has investigated the behavioral effects of intense immediate emotions occurring simultaneously in large numbers of people, in particular in relation to inequality, social movements, and stratification (Barbalet 1998; Flam 1990, 1998; Goodwin *et al.* 2004; Neckel 1999; Summers-Effler 2002).

In the following paragraphs I will focus on the *preliminary stages* of such collective emotional effects. I am interested in how the social structuration of emotions and of (sometimes subliminal) basic affective reactions influences and ultimately structures everyday social action, in particular decision-making in social and personal contexts. These are of particular interest because of their ubiquity in the social world and because they are often characterized by complexity, uncertainty, indeterminacy, and incomparability of consequences.

As the review of the literature on emotion and decision-making has shown, the influential character of emotions is not supposed to be arbitrary, but rather systematic and guided by certain rules and patterns. The somatic marker hypothesis even suggests a causal relationship in that physiological reactions in some cases necessarily precede decision-making in situations of uncertainty and indifference. Elster (1999, p. 288 f.) has emphasized that, strictly speaking, such situations are only seldom characterized by true indifference, but instead by an incomparability of decision-outcomes. As an example, Elster (1999) mentions the case of someone who wants to buy a car, but is unable to decide between two identically priced models. If this customer was truly indifferent, i.e. indeed had no preference for one of the two alternatives, then a price reduction of just one cent should be the deciding factor. Empirically, however, this is hardly ever the case.

The role of emotions in decision problems characterized by incomparability or immeasurability of consequences has been aptly described by Evans's (2002) "search hypothesis of emotion":

> So even for a simple decision like arranging an appointment with the doctor, the set of possible outcomes for each action is in principle unbounded. Therefore, listing the possible outcomes of any given action is a potentially endless task. Yet, if I am to make a decision, I must stop listing outcomes at some point . . . At some point, you must stop thinking, and start acting.
>
> (p. 499)

De Sousa (2010) makes a similar argument:

> For the number of goals that it is logically possible to posit at any particular time is virtually infinite, and the number of possible strategies that might be employed in pursuit of them is orders of magnitude larger. Moreover, in considering possible strategies, the number of consequences of any one strategy is again infinite, so that unless some drastic pre-selection can be effected among the alternatives their evaluation could never be completed. This gives rise to what is known among cognitive scientists as the "Frame Problem": in deciding among any range of possible actions, most of the consequences of each must be eliminated a priori, i.e. without wasting any time on verifying that they are indeed irrelevant. That this is not as much of a problem for people as it is for machines may well be due to our capacity for emotions.
>
> (§8)

In situations to which the frame problem applies, emotions seem to play a crucial role as interrupts to an otherwise endless iteration of comparisons. In rational choice theory, the concept of "bounded rationality" captures this function of emotions. According to this notion, although people tend to look for an *optimal* solution, they proceed in a stepwise fashion and usually opt for the *first best* solution (a strategy known as *satisficing*). This would convert the frame problem into a "search problem" consisting of finding a strategy that looks for possible solutions and at the same time checks whether a solution is adequate or not. Here, too, emotions are supportive in serving as informational cues in the checking of the adequacy of solutions (Evans 2002, p. 502 f.).

This rationality-supporting view of emotions and more basic affective reactions or "gut feelings" is criticized by Elster (1999). He claims that a person acting rationally knows anyway when to use affect-driven heuristics and rules of thumb and when an in-depth deliberative analysis is more appropriate. Consequently, Elster finds misleading the notion of emotions as supplements for rationality, bridging the gap between instinct-like reflexes and perfect rationality. Instead, he suggests that emotions are a functional equivalent to those rational capacities whose activity they at times disrupt: "The emotions do solve problems—but problems that are to some extent of their own making. The capacity of the emotions to supplement and enhance rationality would not exist if they did not also undermine it" (Elster 1999, p. 291).

This criticism is in fact fueled by two different issues: the question of what emotions *are* (i.e., how they are defined) and what they *do* (i.e., what functions they have) (see also Evans 2002). Elster's criticism is mostly based on an answer to the first question that defines emotions as a counterpart to rationality (Elster 2004, p. 33). If the role of emotion in decision-making is approached from this vantage point and they are seen as either "sand in the machinery of action" (Elster 1999, p. 284) or as supporting and supplementing rationality, then one inevitably runs the risk of always setting up emotions as a sort of straw man whenever reaching the limits in explaining human action. Emotions then become nothing more than a semantic placeholder for cognitive strategies (such as heuristics) that have not yet been satisfactorily understood.

If, however, the question of defining emotions is answered differently—as it has been done in the preceding chapters—then the role of emotions in decision-making can be fruitfully approached from a functional perspective. One of the functions of emotion has clearly been illustrated by research related to the somatic marker hypothesis. According to this account, emotions promote decision-making through basic affective and physiological reactions.

Elster (1999) argues that this function can in principle be fulfilled equally well by other mechanisms, such as certain rules of thumb and other simple heuristics that have been shown to lead to equally good or even better decision outcomes than reliance on "gut feelings" (Elster 1999, p. 295). From this, Elster concludes that the influence of emotions on decision-making tends to be arbitrary rather than systematic. This would render their crucial function in stimulus–reinforcer learning (as discussed in Chapter 2) unlikely, since reinforcers have to set in immediately and regularly after a particular action is chosen, which also should occur on a regular basis. Elster (1999) mentions a number of examples that do resemble typical situations used in studies on somatic markers but which, according to Elster, can be solved without any emotional involvement, for instance estimating time intervals or evaluating share prices.

However, Elster's discussion of the role of emotions in decision-making (in particular regarding somatic markers and the frame problem) disregards two important aspects: First, the somatic marker hypothesis in its original version explicitly focuses on *personal* and *social* decision-making domains. As such, it is strongly related to implicit knowledge and autobiographical memory and cannot readily be compared with the semantic and declarative knowledge that is central to virtually all of Elster's counterexamples. Second, Elster's critique tends to disregard the close link between somatic markers and past experiences and situations. Elster expresses concerns that emotions are all too often taken as a straw man that obscures the non-affective mechanisms in decision-making, such as rules of thumb or other heuristics. From a functional view, however, it is quite clear that emotions do have a role in decision-making and social action that cannot be substituted for by alternative and purely cognitive mechanisms. This is because basic affective reactions and full-blown emotions provide a (at times non-conscious, pre-reflexive) link between an immediate decision-making situation and past (affective) experiences in comparable situations.

It is precisely this link that constitutes the second key element of the present investigation. It is fundamentally based on the assumption of the social structuring of emotions and—in personal and social domains—extends and projects these structures of feeling into the realm of everyday decision-making and social action. The main thesis is thus that social action in many domains of social life exhibits patterns and regularities that mirror those related to the experience of affect and emotion. This *affective action* therefore constitutes the second element in our understanding of emotions as a bi-directional mediator between social structure and social action.

Summary

As I will show in detail in the following chapter, one cannot always draw clear distinctions between social action and emotion, primarily because of emotions' inherent expressive components and associated behavioral tendencies. At the level of emotion expression the structures of feeling have clear and direct repercussions for the social environment, in particular through face-to-face interaction. But it is equally clear that understanding emotions solely on the basis of Weber's classical conception forgoes much explanatory potential inherent to emotion.

Looking jointly at both the direct effects of emotions on decision-making outline in the preceding paragraphs and at the indirect effects illustrated earlier, this potential becomes even more compelling. On the one hand, this is because the affective influences on cognitive information processing underlying most attempts at rational decision-making are not as arbitrary as conventional wisdom would have it. Quite to the contrary, the available evidence suggests that the influence of emotion is systematic in the sense that:

- it facilitates the mood-congruent recall of information which is crucial to decision-making;
- subjective feelings are taken into account as *information* in decision-making just like any other kind of information; and
- it prompts different modes of information processing which are decisive to decision-making.

Thus, there are systematic influences both on the *contents* of decision-making (i.e., on the information that is considered) and on the *processing* of information in decision-making (heuristic vs. analytical). The indirect effects of moods and emotions mostly depend on the valence of an existing feeling and decisively determine the mode of information processing. Positive affect, signaling unproblematic situations, gives rise to case-based and schematic information processing that rests on pre-existing and verified knowledge structures and makes use of heuristics. Negative affect, signaling problematic or unconventional situations, prompts more detailed and in-depth processing.

Furthermore, information is used selectively in decision-making based on past and present emotions. In terms of memories, it is "filtered" by mood-congruent recall. In terms of newly acquired information, it is filtered by existing stocks of knowledge. Thus, the available information hardly conforms to the rationality criteria outlined above. In addition, direct influences of present emotions (e.g., through somatic markers) further bias decision-making based on past experiences, whether positive or negative.

Given the strong arguments for a bi-directional relationship between emotion and social structure and the mediating role of affective influences on social action therein, it does *not* necessarily follow that socially structured emotions always and at any time promote actions that are structure-reinforcing or -reproducing. Affective influences on social action may both strengthen and weaken existing social structures (or reproduce as well as undermine social order), primarily because society is not a homogeneous unit or "container" and is characterized by, for example, cultural diversity and intersectionality (e.g., Crenshaw 1991). However, given that there is widely shared societal consensus regarding the denotative and affective meanings of behaviors, identities, settings, objects, and other concepts (Heise 2010), and because social institutions exist to promote convergence in these meanings (for example, through the legal and educational systems), one may assume that the vast majority of affective reactions are widely shared within society and thus structure-reproducing.

Given the research outlined so far, it is plausible that positive affective reactions generally contribute to the reproduction of existing social structures, primarily because they are indicative of actors concurring in their appraisals of structural and symbolic features of the social environment. Also, general knowledge structures, representing shared socialization experiences and social learning in institutionalized settings, are seldom called into question and are part of what is "taken for granted." Similarly, positive mood-congruent recall may facilitate a generally more positive perception of one's environment. In concert, these interactions may constitute a self-reinforcing feedback loop that in cases of positive emotions facilitates actions that contribute to the reproduction of existing social structures.

Conversely, it is plausible that negative emotions tend to weaken existing social structures because they encourage individuals to deviate from established and habitualized courses of action (see also Turner 2011). In negative moods, newly acquired information is analyzed in greater detail and is less affected and filtered by general knowledge structures. Negative emotions also increase the likelihood that existing knowledge is updated and reconfigured, leading to actions and behaviors that oppose existing structural characteristics of the social environment. In line with this reasoning, for example, Barbalet (1998) has emphasized the emotions of envy and resentment, Neckel (1991, 1999) stresses the importance shame, envy, and rage, and Scheff (2000) highlights anger and shame. Also, the literature on social movements makes frequent reference to negative emotions as a driving force of social change (cf. Flam 1998; Goodwin *et al.* 2004). However, negative emotions can also contribute to the maintenance of structures when leading to inactivity, as is possible in the case of anxiety and insecurity. The following chapter extends these conjectures and adds a third argument to the main thesis by investigating the effects of socially structured emotions in face-to-face encounters.

4   The affective structure of social interaction

The social causes and effects of emotions and their neural and cognitive underpinnings discussed in the previous chapters are an indispensable part in arguing for emotions as bi-directional links between social structure and social action. However, such an argument remains incomplete as long as it does not show how emotions are involved in wider social structural dynamics beyond their influence on thinking and social action. To this end, it is promising to look the immediate social context in which social action is embedded and to investigate the role of emotion in social interactions, from which larger dynamics may evolve. This chapter therefore shifts the focus from the mechanisms of emotion elicitation and affective action that are "internal" to actors towards an understanding of emotions as constitutive components of any social encounter and situation.

Social interaction is a key element in sociological theory in a number of different ways, but primarily as a locus for the exchange of significant symbols, the negotiation of meaning, the presentation of self, and, more generally, the constitution of intersubjectivity (Turner, J. H., 2002). It is noteworthy that two well-established theories of the emergence and reproduction of social order are at the same time prominent sociological emotion theories, namely studies by Randall Collins and Jonathan Turner. Both emphasize the fundamentally ritual and patterned nature of social interaction and consider it a nucleus for the evolution of larger structural dynamics. In their view, patterns of social interaction are primarily brought about by the exchange of emotional (and other) resources, based on the seeking of emotional gratification (Collins 2004; Turner & Collins 1989). Looking at this and other possible models of aggregation, one could argue that the social mechanisms of emotion elicitation and affective action can easily be scaled up to large-scale structural dynamics. However, an essential characteristic of affective and emotional reactions is their—often involuntary—expressive component, which considerably corroborates their potential in explaining social order.

Although Jonathan Turner (2002) does include emotion expressions in his theory of emotion, emphasizing the importance of basic emotions and their supposedly universal expressive patterns, he hardly relates them to his arguments on social structuration. Many social constructionist theories of emotion have set quite opposite priorities: Although they are not concerned with face-to-face situations as nuclei of structural dynamics, they are interested in how emotion expressions reflect and reproduce social order on a symbolic basis, primarily through norms and rules that constrain emotional expression (Hochschild 1979; Thoits 2004). These two areas of interest well reflect the traditional opposing views of "positivists" and "constructionists" on what emotions are and how they are socially and culturally constructed.

Aside from these debates, there can be no doubt that emotions are closely associated with different kinds of bodily signals such as facial expressions, voice pitch, gestures, perspiration, changes in complexion and posture. In this way and through verbal language, emotions become essential elements of social situations, as Zajonc (1980, p. 153) noted: "Affect dominates social interaction, and it is the major currency in which social intercourse is transacted." In most cases, people tend to verbalize their emotional experience, such as when swearing at others in traffic, when feeling mistreated, or when football fans yell out their anger or joy. But what would these behaviors be without the non-verbal expressive signals associated with the emotions, which seem to be the emotional "spice" in social interaction?

It is not without reason that conveying emotions without bodily non-verbal signals is considered an art, for example in literature, music, and the fine arts. The bodily, physiological manifestations of emotion are so important in social interaction because they can be voluntarily controlled only to a certain extent. Correspondingly, the various facets of emotional expression are often regarded as indicators of "authenticity." It also frequently happens that one is asked about one's emotional state without actually being aware of any acute and dominant feeling or emotion—let alone an expression thereof. Only after giving it some thought does one (possibly) come to the conclusion that one is indeed in a particular sad or happy mood.

The poker face may be instructive here. The term refers to those gifted players who are able to suppress even very subtle emotion expressions, so that other players are unable to make inferences about current feelings or underlying cognitive appraisals based on expressive behavior. At the same time, those players are often able to decipher subtle expressive signs in other players and use this information to their own advantage. This affective dimension of intersubjectivity, though not necessarily conscious or based on symbolic gestures, crucially facilitates the reciprocal attribution of situational appraisals and action tendencies.

This emotion-based understanding of others enables actors in social interactions to attribute beliefs, desires, and intentions to one another. This way, the bodily manifestations of affective (and cognitive) states become important cues guiding and facilitating social interaction. Moreover, these cues can in turn be appraised by co-present actors and elicit further emotional reactions (such as embarrassment resulting from seeing someone who is ashamed).

But what is the role of emotion expressions in the emergence and reproduction of social order? This chapter develops the main argument that the social structuration of affect and emotion is nothing "private" or "personal," but instead almost always disperses into the immediate social environment and can be accessed, appraised, and experienced by co-present others. This way, patterns of emotional responding and affective action are (even if only involuntarily) disseminated and communicated to other actors and may establish or corroborate the social structures of feeling.

To better understand these processes, it is crucial to look into the basic mechanisms of the *encoding* and *decoding* of emotion expressions. Furthermore, the emotion expressions of one actor often serve as emotion-eliciting cues in another individual. Resulting emotions may either be complementary (as in the case of "emotional contagion," see below) or contradictory. For example, if I observe an angry facial expression, I may almost automatically be "infected" with anger (for example, during a protest march), or react with fear of being harmed, depending on the social context. In that sense, it will be important to investigate the functions of emotion expression, both at symbolic and pre-reflexive levels.

Aside from that, sociological research has consistently emphasized the importance of social norms in both experiencing and expressing emotions in specific situations (Hochschild 1983; von Scheve, 2012a). Thus, emotions— like any other behavior—can be socially appropriate or inappropriate. To adapt emotions and their expression to social expectations, actors employ various strategies of emotion work or emotion regulation. Hence, as the second part of this chapter shows, social norms related to emotions are a decisive ingredient of *social control*. Although these norms first and foremost target emotions, they implicitly also target deviant forms of affective action, i.e. forms that are incompatible with the social order.

Importantly, the relationship between social norms and emotions can also be conceptualized from the opposite perspective, focusing on the question of what role emotions play in the emergence of and compliance with all sorts of social norms as a critical element of social order. Investigating this question in the final part of this chapter will give insights into the fundamental characteristics of social norms and their potential to guide social action. Likewise, the final part will discuss the functions of social norms that support the reproduction of social order in interactions, in particular through cooperation and coordination, and their affective basis.

The expression and communication of emotion[1]

Integrating the facial expression of emotion into sociological models of social interaction requires clarity about the characteristics and qualities of facial expression. With respect to explanations of social order, and in particular micro-social interactive orders, the relevant qualities are above all those indicating linkages between the social or cultural embeddedness and expressive behavior and inform us about the status of facial expression in comparison to other forms of behavior, such as social action. In this light, research on the facial expression of emotion can be divided into two largely antagonistic standpoints oscillating between conceptions of expressive behavior as either biologically hardwired or socially constructed (Russell 1995; Russell *et al.* 2003). Sociological, anthropological, and social psychological approaches tend to focus on social norms and practices regulating facial expression and adapting it to specific social demands (Thoits 1990; Zaalberg *et al.* 2004). Physiological approaches often emphasize the biological causes and the automatic and involuntary nature of facial expression (Ekman & Friesen 1975; Tracey & Matsumoto 2008).

These latter accounts are generally based on the hypothesis of the universality of facial expression, whose origins lie in evolutionary approaches to (basic) emotions (Tomkins 1962) and corresponding cross-cultural studies (Ekman & Friesen 1975). The hypothesis of universality is essentially based on two assumptions. First, it assumes that specific patterns of facial expression are the biologically determined consequence of an underlying discrete emotion. Second, its supporters postulate that both these specific patterns of facial expression and the ability to make reliable inferences about the underlying discrete emotions are cultural universals. These assumptions also imply that the number of distinct facial expressions of emotion is in principle limited to the number of underlying discrete emotions—a view found especially in concepts of basic emotions (cf. Ortony & Turner 1990).

In the sociology of emotion, it is mostly Turner (2002, 2007) who relies on these supposed qualities of emotions and their expression. He assumes that (basic) emotions are always, and for the most part involuntarily, represented and communicated by a universal facial code that allows other actors to clearly infer the underlying emotional states, situational appraisals, and action tendencies. Though the assumption that expressions are largely universal has had a decisive influence on empirical research over the past decades, it has not gone uncontested. Social constructionist research in particular has under-lined the fact that not only possible universals are relevant to the analysis of facial expression, but above all social norms and practices that subtly adapt expressive behavior to prevailing social conditions (Russell 1994, 1995). Apart from these implicit normative influences, the expression of emotion may also be deliberately intensified, suppressed, or simulated for various (instru-mental) reasons. This view is well known in the sociology of emotion and is reflected in such key concepts as "emotion work" and "emotion management" (Hochschild 1983; Thoits 1990).

But even supporters of the universality hypothesis do not deny this possibility in principle, as is evident in the concept of "display rules" (Ekman 1972, p. 225). Generally, though, they assume the existence of a universal emotional response that clearly precedes any attempts at regulation. In this vein, they tend to dispute that even the primary genesis of facial expression is significantly influenced by socio-cultural factors (Matsumoto *et al.* 2008). Early sociologists of emotion, meanwhile, convincingly argued that such factors do in fact play a critical role. Since then, however, few attempts have been made to demonstrate the influence of such factors empirically. As a result, even sociologists of emotion sometimes (uncritically) adopt the prevalent view that facial expressions are universal (e.g., Turner 2002, p.86). More recent studies, however, indicate that facial expression is highly dependent on social contexts, both with respect to the long-term, persistent molding of expressive behavior and short-term adaptation in specific situations (see Zaalberg *et al.* 2004). These studies suggest revising the hitherto dominant and dichotomous conception of universals vs. social constructs in facial expression in favor of a model that avoids marginalizing the evolutionary and corporeal specifics of emotional expression by accounting for processes of social adaptation also on the level of physiology. Supplementing Jonathan Turner's (2002, 2007) approach with these assumptions shows that the role of emotions in social interaction goes far beyond the mere signaling of transactional needs and entails a *sui generis* potential for the emergence of social order.

The encoding of facial expression

Facially expressive behavior occurs in a wide variety of situations and under varying conditions, not only as the result of an emotion, but also as an intentional communicative sign, as the expression of cognitive activity (e.g., furrowed brows during intense concentration), or as a learned symbol such as raised eyebrows (Keltner *et al.* 2003, p. 418 f.). The fact that we ascribe specific emotions to actors solely on the basis of their facial expression has not only led to the everyday theory that such expressive behavior is the unambiguous expression of an underlying discrete emotion, but has also resulted in a remarkable state of affairs within the scientific debate: "Surprisingly few studies have tested the basic claim of EEs [expressions of emotion]: Emotions cause them" (Russell *et al.* 2003, p. 336). That expressions of emotion are in fact caused by an underlying emotional state is constitutive of the universality hypothesis (see Ekman 1972, 1982). In this view, specific patterns of facial expression are part of biologically hardwired configurations of certain components of an emotion and are triggered automatically upon emotion elicitation. According to this model, facial expressions allow unequivocal and universal inferences about underlying emotional states (Carroll & Russell 1996; Parkinson 1995).

Scholars holding this view generally rely on studies demonstrating correlations between a particular type of expressive behavior, subjective experience, and other physiological indicators, such as increased heart rate or peripheral blood flow (Levenson 2003). Cross-cultural studies also seem to confirm this perspective; individuals in different cultures exhibit identical facial expressions when experiencing the "same" emotion lend support to the thesis that emotions are, first, the cause of specific expressions and, second, linked to them biologically. In line with this, Keltner and colleagues (2003, p. 419 f.) conclude that we may also consider facial expressions as reliable indicators of emotional states cross-culturally.

As clear as these connections between emotion and facial expression may appear, they become problematic on closer examination. A smile, goose bumps, and a reddish complexion are neither necessary nor sufficient conditions for joy, fear, or shame. Critics of the universality hypothesis therefore claim that it is not biological predisposition but above all the social context that is crucial to which expression is actually displayed (Kappas 2002a). For example, Russell and colleagues (2003) cite studies showing that winners of Olympic gold medals smile particularly often only when involved in social interaction. Children smile just as often whether they have mastered a task particularly well or failed at it—the sole modifying factor is the presence of other actors (Holodynski & Friedlmeier 2005). Alongside whatever subjective feelings actors may experience, the frequency and intensity of smiling when watching TV comedies is also highly dependent on the particular social context and their relationship with co-present actors (Hess *et al.* 1995).

These "audience effects" largely tally with behaviorist–evolutionary models of expression. Fridlund (1994), for example, argues that facially expressive behavior has developed over the course of evolution primarily as an instrument of social interaction. For him, it chiefly serves the intentions and motives of the transmitter, by communicating action requests as well as intentions or information about the status of an interaction. This interactional focus of expressive behavior is also backed up by studies that were unable to demonstrate linkages between subjective feeling and facial expression in many situations and instead found that expressive behavior occurs significantly more often in social than in non-social situations (see Hess *et al.* 1995). In line with these findings, sociology has highlighted the social and cultural determinants of facial expression, though this is mainly discussed in connection with the (intentional) regulation of expressive behavior (Heise & Calhan 1995). However, the primary encoding of facial expression is unlikely to be systematically modulated solely through intentional attempts at norm-based regulation, as suggested by concepts such as emotion work (Hochschild 1983). In principle, the potential to consciously influence the encoding of emotional expression is limited (see Baumeister *et al.* 2007).

Taking an overall look at the persuasive criticisms regarding biologically hardwired linkages between emotion and facial expression, it is hardly possible to further endorse the universality hypothesis (Kappas 2002a, p. 229). But the reverse conclusion, that (automatic, involuntary) facially expressive behavior is in no way linked with emotions, seems equally implausible: "Of course, there is some association between EEs [expressions of emotion] and emotion; the question is the nature of that association" (Russell *et al.* 2003, p. 341). One way of explaining this connection emerges by drawing on the definition of emotion as a multi-component process as outlined at the beginning. According to this view, facial expressions do not necessarily reflect discrete emotions, but may in principle also occur in non-emotional situations. In cases of emotion as a cause, however, such behavior usually goes hand-in-hand with other components of an emotion, namely physiological reactions, subjective feeling, and cognitive appraisals (Scherer 2005).

In any case, the assumption that the link between emotion and facial expression is biologically fixed not only squanders sociological explanatory potential, but also conflicts with a number of empirical findings. Sociologically, the crucial benefit of taking bodily, physiological processes into account is that they are key to understanding unconscious and involuntary behavior that is a major part of social interaction. Interestingly, sociological theory often assumes that unconscious and automatic behavior cannot be subject to systematic social molding or "construction" and is thus of limited sociological explanatory value. The findings outlined above, however, suggest that the connections between emotion and facial expression are not as rigid as frequently presumed. Instead, theory and empirical evidence suggest that expressive behavior is highly context-specific and that the link between context and expression is the result of social practices, conventions, and the internalization of emotion norms. There is no reason to believe that this specificity can be put down solely to intentional, norm-oriented regulatory strategies. Rather, it is also present within automatic expressive behavior.

For sociological arguments on the structuration of social interaction, this view begs the question of the social consequences of this "plasticity" of the encoding of facial expression. If actors are socialized in stable social structural and institutional arrangements and exposed to similar symbolic and normative orders, one may assume that their expressive behavior is somehow attuned or "calibrated" to these conditions. In face-to-face interactions, calibrated facial expressions may function as indicators of membership in a social unit, for instance as a component of a *habitus*. Within social units, the efficiency and accuracy of calibrated expressions should approximate that of biologically hardwired expressive patterns. In light of this, the question arises on which levels of social differentiation distinct "patterns" of expressive calibration emerge, for example on the basis of gender, ethnicity, social class or milieu, or larger units, such as nation states. Furthermore, it is important to identify the possible consequences of the calibration of facial expression for inter-actions involving actors from different social units. In what follows, I will explore the hypothesis that the social calibration of facial expression, especially in terms of automatic and involuntary expression, is a crucial factor in the reproduction of existing social order, because the closer actors are to one another within the social space, the more smoothly the recognition and inter-pretation (decoding) of facial expression proceeds.

The decoding of facial expression

The question of whether facial expressions can be decoded (recognized, interpreted) in the same way in different social and cultural contexts has prompted a number of comparative cross-cultural studies since the 1970s. In particular, Ekman and colleagues (Ekman 1972; Ekman & Friesen 1975) examined this question by having subjects look at photographs meant to show (Westerners') prototypical expressions of basic emotions. Results show that the probability of recognizing these expressions (by verbally labeling them) in different cultural contexts is fairly high. Much the same applies to other components of emotion, such as eliciting events, cognitive appraisal, and action tendencies (Keltner *et al.* 2003; Russell 1994). However, these studies have been criticized not only from a methodological standpoint but also with regard to their far-reaching conclusions, which are crucial to the universality hypothesis (see Elfenbein & Ambady 2002; Ekman 1992b; Haidt & Keltner 1999; Russell 1994).

It is largely uncontested that facial expression correlates with other components of an emotion. But that facial expression in a decontextualized and artificially created situation, as with photographs, is sufficient to reliably infer a corresponding, underlying emotional state is debated. Doubts have primarily been expressed about the thesis of "facial dominance," according to which the decoding of facial expression and the attribution of discrete emotions may not only occur independently of contextual information, but is successful even when facial and contextual information are contradictory (Ekman 1972; Izard 1971). According to the critics, the ability to reliably recognize emotion expressions requires both shared experiential knowledge as well as situation-specific contextual information. Ekman and colleagues have explained the comparatively high variation in some of their data with reference to display and decoding rules, i.e. norms that regulate expressions and also lead to commensurate expectations regarding their decoding (Buck 1984; Ekman 1972). In this view, social norms are also the primary cause of cross-cultural differences in inferring discrete emotions from facial expressions (Keltner *et al.* 2003, p. 421).

One possible explanation for the fact that cultural universality and facial dominance are still prominent paradigms in emotion research, despite the inconsistencies and critique, is that existing studies have mainly focused on identifying similarities between cultures, not differences (Elfenbein *et al.* 2002). More recent meta-analyses, however, paint a markedly more differentiated picture and put forward a number of alternative, sociologically more relevant explanatory concepts (see Elfenbein *et al.* 2002, Hess & Thibault 2009; Russell 1994, 1995): a "gradient" in recognizing facial expressions (Haidt & Keltner 1999); "minimal universality" and limited "situational dominance" (Carroll & Russell 1996; Russell & Fernandez-Dols 1997); and an "in-group advantage" in emotion recognition (Elfenbein & Ambady 2003a, 2003b; Elfenbein, Beaupré *et al.* 2007).

While they do not necessarily place less importance on certain universals in expressive behavior, these perspectives view universals merely as specific configurations of the micro-components of facial expression (e.g., specific combinations of activity in the facial musculature) and not as invariable elements of a discrete (basic) emotion. These models thus attribute a much greater flexibility to facial expressions, though without abandoning the idea that a given expression has a certain universal core. In view of explaining the emergence and reproduction of micro-social order, however, these approaches throw up two key questions: First, what are the moderating factors that bring about this presumed systematic variation (i.e., calibration)? And second, which mechanisms ensure that expressions can be reliably decoded, at least within pre-existing systems of social order (i.e., within stable social units)?

There is much to suggest that not only culture in an anthropological or "localist" sense, but also intra-societal cultural differences, such as ethnicity, gender, socio-economic status or other indicators of social inequality, play a long-term moderating role (Elfenbein, Beaupré *et al.* 2007). Aside from the present emotional state of actors decoding others' expressive behavior, the situational context is a further short-term moderator of decoding accuracy: Whether a smile is interpreted as embarrassed or polite depends not only on the facial expression but also on situational information, especially on shared implicit and explicit knowledge about these situations, in particular familiarity with relevant decoding rules. If actors lack suitable situational schemas and are unaware of their possible implications for emotional expression, contextual information can help appropriately decode expressions only to a limited degree.

With respect to the mechanisms leading to systematic differences in the ability to decode facial expressions, the supposed "gradient" of recognition is comparatively conservative. This view rests on the assumption that only expressions of certain emotions can be decoded universally (Haidt & Keltner 1999). It is, however, difficult to define these emotions, as they vary depending on the methods used. According to the proponents of the recognition gradient, at least anger, disgust, happiness, surprise, sadness, fear, and embarrassment can be reliably decoded cross-culturally. Shame, contempt, pity, and pleasure are often interpreted differently (Haidt & Keltner 1999, p. 257). The concept of "minimal universality" seems much better suited to explaining the social differentiation and calibration of expressive behavior (Russell & Fernandez-Dols 1997). Here, only certain components of an expression are considered universal, such as up- or down-turned corners of the mouth, open or closed eyes, and straight or raised eyebrows. These components represent a necessary and, in connection with contextual information, sufficient condition for the reliable recognition of a discrete emotion, which suggests "situational dominance" as opposed to facial dominance (Carroll & Russell 1996; Russell 1995, p. 382 f.; Russell *et al.* 2003). The third alternative also clearly relates to the social differentiation of expressive behavior. In a meta-analysis of 165 cross-cultural studies, Elfenbein and Ambady (2002, 2003b) show that the sociocultural proximity of encoders and decoders is crucial to the successful decoding of expressions. The greater the sociocultural "fit," the greater not only decoding accuracy but also decoding speed. This "in-group" advantage in emotion recognition obviously diminishes the more cultures come to resemble one another, for example through spatial proximity or a high degree of communication (Elfenbein and Ambady 2002, 2003b). The in-group effect has also been demonstrated in smaller and less stable social units such as work teams (Elfenbein, Polzer *et al.* 2007).

The last two approaches in particular may serve as sociologically relevant explanations of systematic variation in the recognition of facial expression. First, they indicate that recognizing expressions exhibits certain universal features, though this allows recognizing discrete emotions only in combination with socially shared stocks of experiential knowledge. It is thus not safe to assume that discrete emotions can, in general, be decoded automatically and independently from situational information. Second, they show that socio-cultural differences in expressive behavior go hand-in-hand with an enhanced decoding accuracy within social units. These approaches thus provide a plaus-ible argument for the existence of basic physiological mechanisms for encoding and decoding expressions, which, importantly, can be systematically modulated to generate expressive "dialects" or "accents," much like in verbal language (Elfenbein, Beaupré *et al.* 2007; Marsh *et al.* 2003). Although these accents differ from one another, within social and cultural units they develop the same precision, robustness, and decoding efficiency of supposed universals.

In summary, these approaches not only combine the explanatory potential of universalist and social constructionist models of emotion expression, but also do much to clarify the hypothesis stated at the outset, namely that, in much the same way as symbolically mediated communication, pre-reflexive expres-sive physiological signs are shaped or calibrated by cultural and social struc-tural factors and thus contribute to the patterning of social interactions and the emergence and reproduction of micro-social orders. In particular, the in-group advantage in emotion recognition indicates that the involuntary and automatic expression of emotion and the ability to decode facial expressions are crucially dependent on the social environment. Most interestingly, this influence also applies to those domains traditionally thought to be determined largely by biological factors.

Neuroscience perspectives on decoding and recognition

Taking a closer look at the bodily aspects of the decoding of emotion expres-sions might further clarify the interactions between biological and socio-cultural principles of decoding. An adequate starting point for this endeavor is the basic principles of visual perception. In addressing the problem of cross-cultural universality in the recognition of emotion expressions, Adolphs (2002a) distinguishes (early) perception from (subsequent) recognition. Basic perceptual processes occur immediately after visual stimulus onset and are primarily based on activity in the sensory (in this case the visual) cortices, which process the characteristics of a stimulus and their configura-tions, such as geometric features. At this stage, for example, two different faces can be distinguished from one another. In contrast to the mere percep-tion of an expression, which is based solely on information inherent to the stimulus, *recognition* requires additional information that is not inherent to that stimulus. This includes, for example, knowledge about the contexts in which a facial expression has previously occurred, about other behaviors that accompany an expression, and about behaviors of other individuals in the social situation.

The characteristics of the visual system basically allow for a conceptualization and categorization of stimuli in two ways, one based on perception, the other on recognition.

> One could categorize stimuli on the basis of their visual appearance or on the basis of what one knows about them. Some findings . . . suggest that the geometric properties of facial expressions may suffice to classify them into categories of basic emotions, whereas some cross-cultural studies in humans have argued that the category of emotion expressed by the face is in the eye (and in the cultural background) of the beholder. Neither the reductionist former position nor the relativism of the latter provides the whole answer: Categories can be shaped both by perception and by recognition, depending on the circumstances. The percept of a facial expression—how it appears—can be seen as distinct from the concept of that expression—what one knows about it.
>
> (Adolphs 2002a, p. 22)

Taking visual processing as a starting point, the question arises when the various kinds of information conveyed by facial expressions, such as age, sex, intentions, motives, or underlying feelings, become available. Some models assume that some of these categories are processed in early perceptual systems by specialized neural circuits (Haxby *et al.* 2000; see Adolphs 2002a). Others argue that facial expressions are only processed at later stages and rely on conceptual recognition (see Bruce & Young 1986). In light of the debates on the universality of recognition, Adolphs suggests that the perception of certain components of facial expressions may occur universally in early perceptual systems and are later associated with different, culture-specific concepts and meanings. This view is further supported by the fact that cross-cultural universality has mainly been observed for simple categorizations of expressions, while cultural differences have been shown in particular for their semantic conceptualization and symbolic representation (see Adolphs 2002a, p. 26).

This perspective is supported by evidence on the neural basis of face perception and the interaction between perception and recognition processes. One possibility to conceive of this interaction is to regard recognition as a component of perception (Adolphs 2002a). According to this view, recognition can be the result of the perceptual categorization of stimuli based on, for example, geometric features. In view of facial expressions, this would mean that no additional knowledge would be required to differentiate and categorize distinct expressive patterns and to identify discrete (basic) emotions from these patterns. All the necessary information to identify discrete emotions would need to be contained in an expression's visual features. This is largely consistent with Ekman's (1972) position on the universality of the facial expression of emotion. On the other hand, it leaves little to no room for conceptual knowledge in the early perceptual "recognition" of emotion.

Some empirical studies indeed indicate that, in principle, the visual characteristics of certain facial stimuli are sufficient to bring about the structures generally employed to categorize emotions. These studies also show that even the early perception of expression proceeds largely categorically, and that individuals are also able to reliably assign similar expressions to a specific emotion category (Adolphs 2002a). These positions lend credit to the view that the categorization of facial emotion expressions may be isomorphic with respect to the semantic structure of emotion concepts, i.e. that "the physical, geometric similarity between different facial expressions reflects the structure of our concepts of the emotions" (Adolphs 2002a, p. 28).

From this point of view, emotion expressions would differ markedly from symbolic communication, for example, language, and could thus be described as "sub-symbolic" and non-propositional, while at the same time supplying precise references to associated stocks of conceptual knowledge. Thus, socio-cultural characteristics of emotion expression would be based on structures that already exist at the level of early visual perception and represent links between those characteristics and the culture-specific concepts of emotions and expressions that have arisen in particular social contexts.

Adolphs (2002a) also stresses the possible links between early perceptual categorization of facial expressions and conceptual emotion knowledge as ways in which the recognition of emotion expressions (in the conventional sense) extends beyond information inherent to a facial expression. In analogy to the elicitation of emotions, associative memory processes might play a crucial role in linking early perceptual categorization with stocks of (conceptual) knowledge. The simultaneous activation and integration of information of different representational formats within a single concept (of an emotion) may thus include not only the visual properties of a stimulus but at the same time additional information with which a stimulus has been paired in past experiences (Adolphs 2002a, p. 29).

Further insights into the (subcortical) neural basis of face perception come from studies on patients with blindsight. Blindsight is a form of blindness in which not the eyes are impaired but rather parts of the visual cortex. Under this condition, certain areas of the visual field are not consciously registered and perceived, although incoming visual information is still processed in other areas of the brain can therefore, under certain conditions, trigger behavioral reactions (see Adolphs 2002a). Studies on patients with blindsight have played an important role in clarifying how precisely the (unconscious) *perception* (as opposed to the conscious recognition) of visual stimuli takes place. Interestingly, blindsighted patients seem to be able to distinguish certain expressions of emotion (presented in the blind area of their visual field) from each other despite the lack of conscious perception of those stimuli. Neuro-imaging and lesion studies on blindsight patients suggest that subcortical structures, especially the amygdala, play a critical role in the ability to distinguish facial expressions, in particular in cases of fear and other negative emotions (Adolphs 2002b).

Although these studies do not warrant the conclusion that the categorization and differentiation of emotion expressions can take place entirely in subcortical brain structures, this form of decoding probably contributes significantly to the rapid and reliable decoding of expressions and possibly also to the explanation of a high level of cross-cultural universality in decoding certain components of emotion expression. This does not yet, however, explain how discrimination and categorization is linked to conceptual and highly culture-specific knowledge and thus the full *recognition* of facial expressions.

Detailed representations of a facial stimulus that are necessary for recognition in a stronger sense and allow for the discrimination of facial expressions are available in early visual cortices about 170 milliseconds after stimulus onset. To recognize discrete emotions, amygdala and OFC link perceptual representations of an expression to three strategies for producing conceptual emotion knowledge (Adolphs 2002a, p. 53): First, through feedback processes to the visual cortices that contribute to modulating and refining the formation of perceptual representations; second, through links between various cortical regions and the hippocampus to retrieve stocks of conceptual knowledge associated with a perceived expression; and third, via motor areas, the hypothalamus (which is crucial to memory), and the brain stem, through which certain components of an affective response to the facial expression are activated. The third strategy is also crucial to another possible way of gathering information about another person's emotional state, namely contagion and simulation, which will be discussed in more detail in the following section.

Social calibration and emotional contagion

Until now, the contribution of facial expressions to the emergence and reproduction of (micro-)social order mainly reflect enduring and situational influences of the sociocultural environment on the encoding and decoding of expressions. In this section, I will augment this understanding by examining the more immediate effects of this influence on social encounters. I will focus on processes that have already been investigated by scholars such as Gustave Le Bon (1896) and Emile Durkheim (1915)—the spontaneous "transfer" of emotions between actors. Alongside the verbal communication of subjective feelings ("emotion sharing," Rimé 2009), it is chiefly involuntary nonverbal processes that authors tend to cite in explaining affective mass phenomena and collective emotions such as Durkheim's "effervescence."

Many of these processes are familiar from everyday experience. Laughing, for example, can be contagious, and we tend to wrinkle our noses automatically when something abhorrent happens to others. In the literature, these and similar phenomena are usually dubbed "emotional contagion" (Hatfield *et al.* 1994), which is defined as the "tendency to automatically mimic and synchronize facial expressions, vocalizations, postures, and movements with those of another person and, consequently, to converge emotionally" (Hatfield *et al.* 1992, p. 153 f.). This definition includes several components of an emotion, such as subjective feeling, physiological responses, and action tendencies (Hatfield *et al.* 1992). It can also adduce a large number of different mechanisms that may be conducive to emotional convergence or synchronization, for example socially shared appraisals or interpretations of an emotionally relevant situation or the "mental simulation" of an emotion (Fischer *et al.* 2004). Importantly, emotional contagion is not limited to dyadic settings, but may affect several actors at once (Hatfield *et al.* 1992).

Emotional contagion has to be distinguished from other forms of emotional reaction in response to the (consciously) perceived emotions of others. Such reactions are generally based on "fully-fledged" processes of emotion elicitation (including various appraisals), for example when someone responds with anger or envy to an acquaintance's joy. In contrast, emotional contagion is not based on fully-fledged cognitive appraisals of an event, but largely represents unconscious facial responses to others' expressive behavior. Subjective phenomenal experiences triggered by emotional contagion are basically influenced by four factors: activity of the central nervous system, which is responsible for processes of unconscious mimicry; afferent feedback from these mimetically triggered facial expressions; interoception, which lets actors make inferences about their own emotional state and the expressions of others; and finally the social context (see Bourgeois & Hess 2008; Hatfield *et al.* 1992, p. 155; Hess *et al.* 1998, p. 511).

In principle, emotional contagion is based on mechanisms comparable to motor mimicry, which is well documented empirically (Chartrand & Bargh 1999). Motor responses, including facial expression, have previously been defined as a key component of an emotion. The "facial feedback" hypothesis postulates that phenomenal feelings experienced in an emotion episode crucially depend on afferent feedback from the facial musculature (Hatfield *et al.* 1994). As a consequence of specific patterns of facial expression, the subjective feeling of joy, for example, differs from experiences of sadness or anger (Hatfield *et al.* 1992, p. 161 f.; Hess *et al.* 1998, p. 511). Given that motor mimicry occurs regularly in response to facial expressions, the facial feedback hypothesis would predict that the (involuntary) imitation of specific patterns of facial behavior triggers corresponding subjective feelings in the imitating actors (see Hess *et al.* 1998, p. 512).

However, the hypothesis has not gone uncontested. Studies on the role of peripheral physiological responses indicate that emotions may be experienced even without these physiological changes. Of primary sociological interest, however, is the fact that peripheral physiological processes triggered by facial mimicry at least foster specific kinds of emotional experience and may thus lead to emotional convergence among actors. Numerous experimental studies have confirmed that emotional contagion occurs as a result of the expression of happiness, sadness, anger, and fear, demonstrating that recipients display corresponding facial expressions and experience convergent feelings. These studies also suggest that actors who mimic perceived facial expressions tend to experience consistent discrete emotions, and not merely diffuse positive or negative affective arousal (see Bourgeois & Hess 2008). Moreover, evidence indicates that the perception of an expression automatically triggers consistent activity of the facial musculature in recipients immediately following the perception (Lundqvist & Dimberg 1995). It has also been shown that facial mimicry even occurs if an expression is perceived without consciously recognizing it (Dimberg & Thunberg 1998). In addition, unconsciously perceived expressions are clearly sufficient to initiate typical physiological processes, which in turn constitute (the necessary but not sufficient) components of the corresponding emotion.

Despite these findings, it is also clear that emotional contagion remains a comparatively basic process in social interaction. In terms of the quality of the phenomenal feeling, it is hardly comparable with fully-fledged emotions triggered by complex appraisals. With regard to the role of facial expression in the emergence and reproduction of micro-social order, it seems nonetheless plausible that emotional contagion plays a similarly adaptive and stabilizing role for larger social units as do emotions on the level of the individual (Hatfield *et al.* 1992, p. 153). This is all the more likely in view of the suggested social calibration of the encoding and decoding of facial expression and the in-group advantage in recognition, especially given that contagion may be viewed as a core physiological component of the decoding of expressive behavior. Though we presently lack empirical evidence, it is reasonable to assume that emotional contagion is likely to occur more efficiently within social and cultural units (to which expressive behavior has been calibrated) than across the boundaries of these units (von Scheve & Ismer 2013).

This also makes emotional contagion a potential candidate for a mechanism capable of disseminating the sociocultural foundations of emotion elicitation between individuals without significant friction losses caused by the symbolic and reflexive communication of emotion. Emotional contagion fosters the intersubjectively shared experience of social situations by making situational interpretations and action tendencies of other co-present actors directly and immediately palpable for those in physical proximity. Through emotional contagion, certain physiological and affective components of social action can be interindividually attuned ("affective resonance") in social interactions largely without actors' conscious involvement.

Summary

In adding to symbolic interactionist theories of emotion it becomes evident that the hitherto underrepresented pre-reflexive and "sub-symbolic" components of emotion, in particular facial expression, are a crucial factor in establishing and reproducing social order. The sociological relevance of expressive behavior is especially evident when not only accounting for universals in facial expression (e.g., J. H. Turner 2007), but when emphasizing their plasticity and social calibration, i.e. their involuntary and embodied attunement to the practices and norms of systems of social order. From an encoding perspective, facial expression rests on relatively invariable biological mechanisms that underpin the automaticity and immediacy of certain expressive components. The configuration of these components and their linkage to certain situational parameters, however, is crucially and systematically influenced by actors' embeddedness into social units and systems of social order, both over the long and short term. Consequently, actors embedded in different social and cultural units exhibit marked nuances in facially expressive behavior, making it an interactively highly salient indicator of social differentiation. In analogy to verbal language, this social calibration of emotion expression gives rise to non-verbal accents or dialects, which become part of a physiologically grounded "emotional *habitus*," much like that proposed by Bourdieu (1984).

Importantly, the decoding of emotion expression is also crucially determined by this linkage. The successful decoding of facial expressions of emotion, in particular of discrete emotions, is dependent on situation-specific information and socially shared experiential knowledge; only in some cases can successful decoding be achieved solely on the basis of an isolated facial expression. Actors' decoding abilities develop in analogy to the social calibration of the encoding of emotion expressions, and the closer encoders and decoders are within the social space and the more their shared experiential knowledge overlaps, the more precise this ability probably is.

Finally, these encoding and decoding qualities systematically impact the potential for emotional contagion, such that emotional contagion is likely to occur more efficiently within social and cultural units (to which encoding and decoding of expressions have been calibrated) than across units. Emotional contagion has been shown to be an important factor in the structuring of social interaction and the reproduction of micro-social order, as it promotes not only the transfer of phenomenal feelings between individuals, but also of patterns of physiological arousal and action tendencies coupled to emotional states. This perspective complements the role of emotional gratification in the structuring of social interaction, as proposed, for example, by Collins (2004) and J. H. Turner (2007), since emotional contagion does not exclusively depend on the satisfaction of needs and desires.

In summing up the multitude of theories and empirical findings on emotionally expressive behavior, what emerges for sociology and the affective foundations of social order is a picture similar to that sketched by Silvan Tomkins already in 1962: "The individual who moves from one class to another or from one society to another is faced with the challenge of learning new 'dialects' of facial language to supplement his knowledge of the more universal grammar of emotion" (p. 216). A system of facial expression that quickly and reliably provides information on the status of an ongoing social interaction and the parties involved, while nonetheless reflecting actors' embeddedness into systems of social order, is not only of crucial importance to micro-sociological analysis. When it comes to explaining the structuring of social interaction and the emergence and reproduction of micro-social order, emotion expressions are relevant not only as signals of actors' internal states and concomitants of social (and emotional) exchange, but above all as a pre-reflexive means of intersubjective understanding—of interaffectivity—as well as an indicator of situational interpretations and action tendencies (see also Reich 2010). Given that situational interpretations generally go hand-in-hand with emotional responses, which in turn result in characteristic facial expressions, we may assume, first, that these expressions can be decoded quickly and automatically and thus allow actors to make mutual inferences of underlying emotional states, cognitive appraisals, and likely action tendencies; and, second, that processes of emotional contagion may lead to the convergence or attunement of emotional states and, consequently, corresponding action tendencies.

This three-stage process substantially increases the probability of the emergence of regular patterns of social interaction and the reproduction of micro-social order. This potential becomes even clearer given that this process occurs more smoothly within than across systems of social order. When analyzing social interaction and the emergence of micro-social order, sociological theory and empirical research can undoubtedly benefit from considering a system of intersubjective understanding that deviates substantially from the widespread focus on the symbolic foundations of interaction, while relating isomorphically to symbolic categories and semantic concepts of emotion, such as emotion norms, which will be discussed in the following section.

Emotion regulation and social control

The different perspectives on emotion outlined in the introductory chapter of this volume all suggest that emotions fulfill a number of adaptive functions. They signal contingencies between the individual and the (social) environment in view of well-being and survival and promote appropriate behavioral responses through patterns of physiological arousal and psychological attunement. Richard Lazarus (1991b) has summarized the often evolutionarily rooted adaptive functions of emotion in stating that they "contain the wisdom of the ages" (p. 820).

However, we know from our own experience that emotions are by no means always adequate and functional. Emotion-driven behavior often is not more than a "best guess" that may even cause severe problems. Such problems for the most part are a consequence of the characteristics of modern societies, in which the "wisdom of ages" is actually not always in keeping with the times. It is not the snake in the grass or the bear in the forest that are the primary triggers of fear today, but rather the loss of one's job, an argument with colleagues, or an imminent examination. In many cases, social norms, conventions, moral obligations, and social comparisons have taken the place of nature's challenges and are both triggers of emotions as well as means of controlling them.

Clearly, emotions decouple stimuli from responses and grant flexibility to our behavior and we do not have to rely exclusively on instincts and reflexes. In many cases, we can thoroughly analyze a situation and subsequently decide on a particular course of action. However, as I have shown in the preceding chapters, emotions also systematically influence our capacities for rational thought and decision-making—for better or worse. Although the concept of affective action suggests that affect-driven behavior not only mirrors the "wisdom of ages," but also reflects culture, socialization, and learning, it is often just a best guess that—much like a rule of thumb—might require adjustment to be effective and socially adequate.

> Physical and social environments have changed out of all recognition from those that shaped our emotions, and technological advances have dramatically magnified the consequences that our emotional responses may have for ourselves and others. An irritable swipe that once scarcely raised a welt, is now translated with the greatest ease into a fatal car accident or gun-related homicide.
>
> (Gross 1999a, p. 558)

Despite their automaticity, there are a number of ways to regulate emotions, either deliberately or automatically (Mauss *et al.* 2008), and adapt them to social or individual requirements. The two preceding chapters have illuminated how systems of social order (such as culture and social structure) constantly and effectively influence and calibrate certain components of an emotion, which can well be understood as a form of cultural regulation or social control. In complementing this view, this section will focus on the (intentional) regulation of an already existing emotional state in order to comply with certain social norms. Social norms often determine which emotions and expressions are expected of which actors in specific situations. At a funeral, for example, one is expected to show grief; at a cocktail party one should look happy, and anyone losing their job is expected to be sad and annoyed. These expectations do reflect the increasing functional and structural differentiation of modern societies as well as the prevailing discourses related to emotional intelligence, emotional competence and the imperative to happiness and well-being (Ehrenreich 2009; Illouz 2008; Neckel 2005).

In this section I will therefore elaborate the conjecture that certain social norms (*emotion norms*) at the level of symbolic and conceptual knowledge support and enforce the social shaping and calibration of emotion that operate at various pre-reflexive and bodily levels, as discussed in the previous chapters. In a number of contexts, social norms give rise to or demand structure-reinforcing emotions, serving as means of emotional social control.

This, however, is only one side of the coin. The interplay of social norms and emotions is also critical from the opposite perspective, namely the influence of emotions on the maintenance and enforcement of (other) social norms. Theory and empirical evidence, in particular from philosophy and behavioral economics, suggest that emotions play a crucial role in norm compliance and maintenance. Taken together, I will argue that emotions are important to establishing and maintaining social control in two ways: On the one hand through their normative status that is established by declarative and prescriptive emotion norms that regulate emotions in much the same way as other norms regulate various kinds of behavior; on the other hand by ensuring the compliance and maintenance with social norms in general, which are key determinants of social action and a primary route to the establishment and reproduction of social order.

Emotion norms

The sociology of emotions has primarily focused on one specific category of social norms, namely emotion norms, or what Hochschild (1979) classically termed "feeling rules." Emotion norms constitute a particular category of norms that explicitly and exclusively relates to emotions and their various constituent components. These norms specify which emotions are regarded as appropriate and expected in specific situations: "Feeling norms indicate the range, duration, intensity, and/or targets of emotions that are appropriate to *feel* in specific situations" (Thoits 2004, p. 360). This sets emotion norms aside from concepts such as "ideal affect" (Tsai 2007), "emotion values" (Eid & Diener 2001; Tsai *et al.* 2006), and other emotion-regulatory goals, in that many of these are desired and valued very generally and independently of specific situations (typically, in Western societies, happiness) (see von Scheve (2012a) and Tamir (2009) for a discussion). Moreover, research has theorized expression or display rules, which do not refer to subjective feelings, but instead only to expressive behavior, guiding "appropriate displays of emotion in given situations" (Thoits 2004, p. 360).

Emotion norms can be further specified in view of social situations. Fiehler (1990) has linked categories of situations with emotion norms through conditional statements. To this end, he introduces the concepts of manifestation rules, correspondence rules, and coding rules (Fiehler 1990, pp. 77 ff.). In his model, emotion norms (or rules) codify which emotion is appropriate in a particular type of situation from the perspective of the emoting actor and which one is expected from the perspective of others involved in the situation. Manifestation rules govern which type of emotion is to be displayed in what way in a particular situation (Fiehler 1990, p. 78). They are similar to Ekman's (1972) "display rules." Correspondence rules govern how to appropriately react to the emotion expressions of other individuals in a situation. They codify the types of feelings and expressions that are considered to be appropriate and socially expected in response to an expression of someone else.

Correspondence rules thus complement emotional contagion on a reflexive and conceptual level and have also been described as "mood-sharing" or "mood-joining" (Denzin 1980, p. 257 f.; Denzin 1984; Fiehler 1990, p. 79). Finally, coding rules govern which behaviors are regarded as manifestations of which emotions and refer not only to facial, but also to linguistic, vocal, gestural, or symbolic forms of expression (Fiehler 1990). Fiehler assumes that in situations governed by various emotion-related norms, actors have to engage in a multi-facetted monitoring of their emotions. They have to identify the specific norms that are valid in a specific situation, monitor their own emotions and expressions as well as the expressions of others.

Little is known, however, about the emergence of emotion norms. One way to think of how emotion norms come into being is to assume that specific and recurring emotional behaviors at some point acquire the status of a *descriptive* norm (e.g., Cialdini 2007). Based on the social shaping and calibration of emotions outlined previously, specific emotional reactions to certain stimuli and in certain situations can become conventionalized and may be perceived as "what is commonly felt and expressed" in these situations. Processes of conformist transmission or "copy the majority" heuristics (Henrich & Boyd 1998) may then establish the spreading of such a descriptive norm. On the other hand, emotion norms are clearly *prescriptive* (e.g., Bicchieri 2005) in the sense that they convey what ought to be felt and expressed and that there are mutual expectations and informal sanctions in cases of non-compliance. As prescriptive norms, emotion norms may be transmitted through social practices, explicit learning, and symbolic codifications, such as in manners books (e.g., Elias 1994). This way, emotion norms not only shape emotions but also reflect prevalent cultural views on emotions, their relative importance, and the socially accepted ways of dealing with them. Thus, they play a crucial role in shaping the "emotional culture" of a social unit (Thoits 2004, p. 362; see also Cancian & Gordon 1988; Reddy 2001; Stearns 1994).

Moreover, social norms are both mirrors and outcomes of social orders. In as much as they are "guides for action" they are also "outcomes of practices" and as such subject to revision and modification (Barbalet 1998, p. 23). Given the social structuration and calibration of feelings and emotions and their involuntary elicitation, emotions that occur regularly in certain classes of situations may promote the establishment of emotion norms, as Thoits (2004) argues:

> recurrent individual experiences tend to become emotional conventions, or norms. People develop expectations about the intensity and duration of grief based on their own and the often observed experiences of others, and these expectations (norms) are passed on to others.
>
> (p. 363)

The main impact of emotion norms on the social structuration of emotions and basic affects is that they act as an (intentional) interactive layer of social control on top of the pre-reflexive and bodily levels of the social shaping of emotion outlined in the previous chapters. Looking at the mechanisms involved in the social structuration of emotions and their consequences for social action, it is evident that they develop a normative power, even if only in a descriptive sense. Prescriptive (or injunctive) emotion norms, however, promote the explicit and deliberate social control of emotions that have been more or less calibrated to a social environment. Importantly, emotion norms can be prepositionally represented and communicated through language, text, mass media, and other sign systems.

Thus, there are good reasons to assume that emotion norms are a crucially involved in the emergence and reproduction of social order (Thoits 2004). Violating established emotion norms does not only indicate "deviant" emotions, but most probably also "deviant" beliefs, desires, attitudes, corresponding appraisals, and likely action tendencies. Emotions and emotion norms can therefore be considered an "early warning system" indicating not only the violation of emotion norms but also the likely violation of other norms by subsequent emotion-driven behavior (e.g., aggression). To adapt emotions to social or personal expectations and to ensure compliance with prevailing emotion norms, actors frequently perform emotion regulation, which I will discuss in detail in the following section.

Emotion regulation and emotion work[2]

Emotion regulation can basically be considered from two different perspectives. On the one hand, it can be seen as a process that involves the more or less intentional regulation (i.e., modulation or transformation) of one's own emotional state based on certain norms, values, and desires or specific situational requirements and expectations. This perspective is known as the "two-factor" model of emotion regulation since it presupposes, first, the generation of an emotion (factor one) and, second, attempts at regulating the existing emotion (factor two) (Campos *et al.* 2004). The concept of "emotion work," which is widely used in sociology and will be discussed in detail in what follows, is a good example of a two-factor model.

The second perspective is reflected in one-factor models of emotion regulation. This view holds that "emotion and regulation are one" (Kappas 2011) and the regulation of emotion is not limited to an actual emotion episode, but rather extends throughout ontogenetic development. In this vein, some have argued that "emotion and emotion-control are part and parcel of the same processes and any scientifically viable theory of emotion must also be a theory of emotion-control" (Kappas 2008, p. 15). Hence, Campos and colleagues (2004) define emotion regulation as the "modification of any process in the system that generates emotion or its manifestation in behavior . . . Regulation takes place at all levels of the emotion process, at all times that the emotion is activated, and is evident even before an emotion is manifested" (p. 380).

The one-factor model of emotion regulation is thus consistent with the main conjecture outlined in this volume, namely that emotions are fundamentally and systematically shaped ("regulated") by the social structures in which actors are embedded. However, this approach may lead to an arbitrary view of the regulation of emotion that, in principle, can no longer be distinguished from its elicitation. I therefore suggest looking at the long-term social and cultural influences on emotion not from the perspective of emotion regulation, but rather from the social structural and cultural perspective taken in the previous chapters.

This mirrors Hochschild's (1979) view, which promoted two possible approaches to the "social ordering of emotive experience" (p. 552). The first is based on the analysis of the "social factors that induce or stimulate primary . . . emotions," whereas the second one is "to study secondary acts performed upon the ongoing non-reflective stream of primary emotive experience" (p. 552). In her now classic studies on the social regulation of emotion, she focuses on the second option and takes a two-factor perspective. She coins the "secondary acts" that are performed on primary emotive experience "emotion work" (p. 552). Originally developed in an investigation of the emotional demands of service-sector employees, emotion work corresponds to the "act of trying to change in degree or quality an emotion or feeling" or simply "to 'work on' an emotion" (p. 561). It is closely related to—and in fact an extension of—Goffman's (1959) ideas on *The Presentation of Self in Everyday Life*. As such, her account is strongly influenced by the principles of symbolic interaction.

Hochschild assumes that emotion work in principle serves two goals: to either evoke or to suppress an emotion. Her account of the processes and mechanisms underlying emotion work is inspired by the ways in which professional actors evoke and shape emotions, and she makes explicit reference to Stanislavski's method acting paradigm (Hochschild 1983) to distinguish two types of emotion regulation: "deep acting" and "surface acting" (Hochschild 1983, p. 48). Here, "deep acting" is mainly used synonymously with "emotion work," meaning the management of a feeling or an emotional state, whereas "surface acting" is limited to modulating only the behavioral expression of an emotion. Surface acting thus equals Goffman's (1959) description of impression management in social interactions.

Empirical research has almost exclusively focused on one specific instance on emotion work, namely "emotional labor." Emotional labor denotes emotion work that is performed in organizational and economic contexts. It does not primarily pursue individual goals, but is rather seen as an instrumental strategy to increase the economic success of an organization. Hochschild's classic study on the emotional labor of flight attendants and employees in debt collection agencies provided an empirical illustration of the concept (Hochschild 1983, pp. 89–161) as does a body of more recent studies in the sociology and psychology of work and organization (e.g., Brief & Weiss 2002; Fineman 2003).

In her studies, Hochschild for the most part adopts a critical stance towards the social, psychological, and physiological consequences of emotional labor and emotion work. According to Hochschild, emotion norms, in particular feeling rules, frequently create dissociations between socially expected and actually experienced emotions. This tension gives rise to "emotional dissonance" (Hochschild 1983), which has to be alleviated by means of emotion work. This, Hochschild suspects, in the long run leads to alienation from one's own feelings. Although Hochschild does not frame her work exclusively as an approach to emotion regulation, but rather locates it within social criticism, it has far-reaching parallels with research on emotion regulation, in particular in psychology (Gross 1999a; Weiss & Brief 2001).

Psychological two-factor models of emotion regulation offer insights into the individual and social determinants and goals as well as the social, cognitive, and affective consequences of regulation, as already emphasized by Hochschild. They are therefore well suited as a guiding framework for extending and specifying sociological research on emotion work (Grandey 2000, p. 98; Grandey & Brauburger 2002). Integrating sociological accounts of emotion work and emotion norms with these models promises a better understanding of the links between the individual and sociocultural dimensions of emotion and emotion regulation and, ultimately, to enhance our understanding of the role of emotion in the emergence and reproduction of social order.

Emotion regulation according to two-factor models is often defined as referring to:

> the processes by which individuals influence which emotions they have, when they have them, and how they experience and express these emotions. Emotion regulatory processes may be automatic or controlled, conscious or unconscious, and may have their effects at one or more points in the emotion generative process.
>
> (Gross 1998, p. 275; italics omitted)

This definition encompasses positive and negative emotions as well as attenuating and amplifying emotions. Also, emotion regulation in this view extends to many components of an emotion, for example expressive behavior, subjective feelings, or bodily reactions (Gross 2002, p. 282).

James Gross's process model of emotion regulation, which is shown in Figure 4.1, identifies five different stages of emotion regulation and distinguishes antecedent-oriented from response-oriented regulation. The model is based on an understanding of emotion that is similar to the view outlined in Chapter 1. Emotion, in this view:

> begins with an evaluation of emotion cues. When attended to and evaluated in certain ways, emotion cues trigger a coordinated set of response tendencies that facilitate adaptive responding. These response tendencies

involve experiential, behavioral, and physiological systems. Response tendencies from each system may be modulated, and it is this modulation that gives final shape to the manifest emotion.

(Gross 1999b, p. 528)

This perspective allows for separate consideration of the various stages of emotion regulation and for relating them to the different phases of emotion elicitation. Gross's model is organized along a time axis representing the individual phases of emotion elicitation. Antecedent-oriented regulation, setting in at the early stages of emotion elicitation, refers to strategies that can become effective even before an emotion has fully manifested. Avoiding unpleasant topics of conversation or leaving a situation that constantly produces anger are examples of antecedent-oriented regulation. Response-oriented regulation, on the other hand, refers to strategies aimed at altering certain components of an existing emotional state, for example suppressing the facial expression of one's anger or down-regulating one's arousal in the case of rage.

Antecedent-oriented regulation encompasses several possibilities: the selection and modification of a situation, the deployment of attention, and the appraisal of the situation (see Gross 2002, p. 282). Situation selection and modification relate to strategies aimed at seeking or avoiding situations in which actors expect certain emotions to occur. If a situation cannot be created or selected, it can possibly be modified in such a way that a desired emotion is experienced or an undesired is no longer experienced. Attentional deployment refers to the strategy of attending to particular (selected) aspects of a situation that change an emotion in the desired way. A related emotion regulation strategy, reappraisal or cognitive change, involves the cognitive re-interpretation of a situation or of a certain feature of the situation. Reappraisal constitutes the (intentional) re-evaluation or modulation of a previous emotion-eliciting cognitive appraisal and imbues the situation with a different meaning that gives rise to changes in the ensuing emotion (see Gross 2002, p. 282 f.).

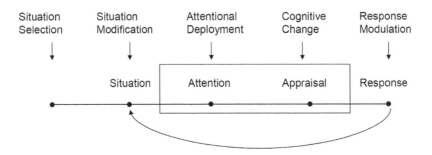

Figure 4.1 Basic process model of emotion regulation.

Source: Reproduced from Gross & Barrett (2011, p. 12). Reproduced by permission of Guilford Press.

The emotion antecedent strategies of attentional deployment and cognitive change—or reappraisal—largely correspond to Hochschild's concept of deep acting or emotion work in a narrower sense (Hochschild 1979; Grandey 2000) (see Figure 4.2). Because Hochschild's work has a focus on emotional labor in organizational settings, it seems obvious that she emphasizes these cognitive regulation processes over situation selection and modification, mostly because employees are limited in their capabilities to select and modify situations. According to Hochschild, emotion work may consist of three elements: cognitive, bodily, and expressive. Cognitive strategies in models of emotion work refer to attempts to "change images, ideas, or thoughts in the service of changing the feelings associated with them" (Hochschild 1979, p. 562). Most interestingly, although hidden in a footnote, Hochschild (p. 562) explicitly relates these cognitive strategies to appraisal theories of emotion, in particular Lazarus's (1968) approach, which are also foundational to psychological process models of emotion regulation. However, emotion work is only seldom seen in the light of appraisal theory. It can be understood as an attempt at "recodifying" situations or at reclassifying them into "previously established mental categories" (p. 562). This deliberate and conscious recodification (reappraisal) acts upon previous automatic codifications and interpretations (appraisals) that gave rise to the initial emotion.

Response-oriented regulation in Gross's model resembles the idea of surface acting in theories of emotion work (see Figure 4.2). Here, Hochschild's (1979, p. 562) ideas of regulating the bodily, i.e. physiological, components or "symptoms" of emotions (e.g., respiratory control) are in line with Gross's view of response-oriented regulation. The same holds for the expressive components that are, strictly speaking, a class of bodily reactions. Importantly, and in contrast to the process model of regulation, Hochschild is interested in bodily and expressive regulation primarily in view of the their effects on the regulation of the underlying feeling, for example trying to smile not only for "interactive" reasons, but also to change the phenomenal feeling (Hochschild 1979). In line

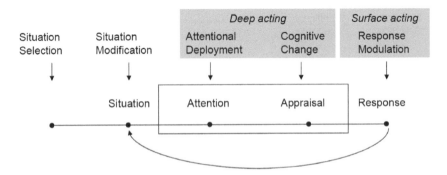

Figure 4.2 Deep acting and surface acting in the process model of emotion regulation.
Source: Gross & Barrett (2011, p. 12) and von Scheve (2012a).

with Gross, she acknowledges that antecedent- and response-oriented strategies often go hand in hand. Importantly, in uncovering the social determinants of emotion regulation, both strategies have to be linked to certain norms and values that serve as emotion-regulatory goals, in particular to the feeling rules outlined above.

Looking at antecedent-oriented regulation, in particular situation selection and modification, it is important to note that these strategies depend on individuals' capacities to actually select and change a situation. This is often neglected in psychological models. These capacities are constrained by several factors, in particular the institutional setting and available resources. Some institutional settings such as third-sector employment with frequent customer contact leave only little room for selecting situations at will. Also, situation selection aiming at emotion regulation in certain areas of the family or in educational settings may be hard to achieve. As a general rule, the more formalized an institutional setting is and the more individuals are bound to a specific social role, the less likely becomes situation selection as a strategy of emotion regulation.

Moreover, selecting and modifying situations requires adequate resources. This includes cultural resources in the broadest sense, such as knowledge on how to change or select a situation, it may require economic resources as a means to actually implement selection or modification, and this strategy may also need the adequate social resources, in particular status and power (e.g., Kemper 1978), that enable individuals vis-à-vis others to change a situation. Importantly, as sociological research has documented over the past decades, these resources are not arbitrarily distributed in society, but highly inter-correlated and associated with social structure (e.g., Massey 2008). Systematic social differences in the available resources to implement certain strategies of emotion regulation should thus—in conjunction with norms and regulatory goals—lead to discernable social patterns in emotion regulation and clarify how regulation contributes to the reproduction of social order.

Aside from these conceptual similarities, linking emotion work to emotion regulation research also offers insights into some of the more debated aspects of Hochschild's theory, namely the concern that emotion work generally gives rise to greater psycho-physiological strains, emotional alienation, and is adverse to health (Grandey 2000; Hochschild 1983). Research on the consequences of emotion regulation has revealed significant differences regarding deep- and surface-acting (or antecedent- and response-oriented regulation). For example, studies have shown that the suppression of emotions can have negative cognitive consequences in view of memory consolidation during regulation (Gross 2002). Interestingly, these negative effects have not been observed for reappraisal and cognitive change. Both strategies seem to be equally successful in actually regulating expressive behavior, but response-oriented strategies lead to more adverse side effects such as increased cardiovascular and electrodermal activity. Also, suppression is less successful than reappraisal in regulating subjective feeling (Gross & John 2002, p. 308 f.).

Besides these individual consequences, there are marked social-interactional consequences of emotion regulation. The negative cognitive and affective consequences of suppression suggest that this strategy is also dysfunctional in social interaction, mainly because it simultaneously restrains both positive and negative emotions. Given that emotion expressions allow inferences about underlying emotions, appraisals, motives, and intentions, then suppressive, response-oriented regulation is likely to cause problems in social interaction. Some studies suggest that, in contrast to reappraisal, suppression in dyadic interactions gives rise to marked physiological changes and stress reactions. Moreover, the suppression of expressions is associated with a reduction in social support, which has proved to be a significant factor in coping with stress and illness (Gross 2002, p. 287). Applying these results to deep and surface acting, it would seem, first, that deep acting holds out better prospects of achieving a regulatory goal than surface acting and, second, that it has considerably fewer negative cognitive and social consequences. However, based on these studies, we cannot draw any conclusions regarding the long-term consequences of deep and surface acting, such as the emotional dissonance and alienation Hochschild suspects.

Looking at the neural correlates of the pre-reflexive and automatic elicitation of emotion outlined in Chapter 2, and having in mind the role of automatic and controlled cognitive appraisals in the generation of emotion, it is interesting to ask how—exactly—reappraisal may change emotions. Assuming that appraisal processes operate continuously and without a fixed start or end point, it seems plausible that the reappraisal of an event may indeed permanently change the event's meaning and, as a result, trigger different kinds of emotions than those based on some "initial" appraisal. If we conceive of appraisals as cognitive processes determining an event's emotional relevance, then they should be subject to executive control like many other cognitive processes. This, however, in some way contradicts assumptions related to the automatic, pre-reflexive, and involuntary elicitation of emotion in the basic (subcortical) affect system. Neuroimaging studies have shed some light on how both domains interact in emotion regulation and Ochsner and colleagues (Ochsner & Gross 2004, 2005, Ochsner *et al.* 2002, p. 1215 f.) have investigated three processes they hypothesized to be involved in cognitive reappraisal:

1. First is a strategy that cognitively reframes an event in an "unemotional" way and keeps this framing in mind as long as the event persists.
2. Second is a process monitoring interferences between top-down re-appraisals that "neutralize" affective reactions and automatic bottom-up appraisals that continuously generate basic affective responses.
3. Third is a process that re-evaluates relationships between internal experiential or physiological states and external stimuli and may be used to monitor changes in one's emotional state while reappraising.

In reviewing a number of studies, Ochsner and colleagues highlight the possible interactions with brain areas implicated in generating basic affective responses and those involved in cognitive control. Interestingly, those areas strongly overlap with those of the basic (limbic) affect system and the circuits related to the processing of more complex and social emotions identified in Chapter 2 (Ochsner *et al.* 2002; Ochsner & Gross 2004; 2005). Reappraisal has been shown to modulate activity in those brain areas that also play a crucial role in the elicitation of basic affective reactions and more complex emotions, for instance the amygdala and the OFC (Ochsner *et al.* 2002, p. 1216; Ochsner & Gross 2004, p. 234 f.). Reappraisal primarily involves activity in several areas of the frontal lobes, in particular the prefrontal, orbitofrontal, and cingulate cortices, which also play an important role in cognitive and executive control, working memory, and action planning (Ochsner & Gross 2005).

In view of the various ways in which emotions can be deliberately regulated and adapted to certain desires and social expectations, two questions arise: First, how can this observation be reconciled with the claim advanced thus far that much of the contribution of emotion to the emergence and reproduction of social order rests on non-conscious and involuntary affective activity? And second, what are the more general implications of emotion regulation for the relationship between individual and society? Regarding the first question, emotions can only be regulated in the way suggested by two-factor models if one is in fact *aware* of them. As emphasized in Chapter 1, subjective feelings are not always perceived as components of an emotion and even though basic affective reactions sometimes occur below the level of awareness they still decisively influence action, behavior, and thinking. This means that only a small proportion of affective reactions (those we are aware of) can be regulated in the way discussed above, i.e. through suppression or reappraisal.

In these cases, and with regard to the second question, the reasons for and goals of emotion regulation become a decisive element in the proposed explanation. Why and to what ends do people engage in emotion regulation at all? Aside from emotion norms and feeling rules discussed above, one of the most straightforward and empirically substantiated answers is: hedonic pleasure (Vastfjall *et al.* 2001). However, there is a broad array of other things that people value that are not necessarily associated with pleasure, for instance, social conformity, health, or utility (e.g., Tamir 2009), that might prompt emotion regulation. Nevertheless, accounts of pleasure and pain as motives of emotion regulation have dominated the literature (Tamir & Mauss 2011) whereas the role of values and goals has been much more at the heart of self-regulation than emotion regulation research.

Generally, understanding the goals and motives of emotion regulation also involves understanding the things that actors *value* in affective and emotional terms. The focus on hedonic pleasure in recent research is rooted in the nature of emotions as intrinsically pleasant or unpleasant experiences. In this vein, the standard view holds that people value and aim at seeking pleasant emotions and at avoiding unpleasant ones. However, it is not completely clear what makes a pleasant emotion pleasant. Research has indeed demonstrated robust cross-cultural differences in the valuation of what Tsai and colleagues term "ideal affect" (Tsai *et al.* 2006). This refers to the affective states that people value, prefer, and ideally want to feel. It is at the core of what a "good feeling" actually is (Tsai 2007). Also, Tamir (2009) has questioned the assumption that people always want to feel "pleasant" emotions. Instead, she highlights the role of short-term and long-term benefits and argues that "unpleasant" emotions, such as anger, are often sought to aide long-term goal attainment. Similarly, research on aesthetic emotions indicates that allegedly unpleasant emotions, for example intense sadness, are often actively sought and enjoyed (e.g., Oliver & Woolley, 2010). Clearly, the regulation of emotion is closely tied to the feelings that are preferred and valued by a person. These values and preferences for certain emotions in certain situations develop in social and cultural contexts and are internalized during the course of socialization. Studies have demonstrated that there are marked differences between cultures in view of which emotions are valued and which are not (Eid & Diener, 2001). Therefore, situation-independent cultural values are as significant as situation-specific emotion norms because they support the assumption of regular patterns in emotion regulation and emotion across large numbers of actors, and thus contribute to the formation of social order.

Interestingly, emotion-related norms and values in recent years have increasingly moved center stage in pubic discourse. Emotions, in contrast to Elias's (1994) assumptions, are more and more becoming a critical and highly codified means of the presentation and staging of the self (Neckel 2005). This is reflected, for example, by the popular scientific self-help and counseling literature on "emotional intelligence" or "emotional competence," suggesting that emotions can always be thoroughly controlled and regulated and be utilized as a means for one's personal goal attainment in an instrumental-rational way (Goleman 1995).

"Making good use" of one's own emotions and seeing them as valuable resources to be cultivated and adapted to one's personal and social circumstances goes far beyond Hochschild's (1983) original view. In contemporary society, emotions are increasingly becoming the principal focus of entire industries, from the mass media and the entertainment industry to the health sector and management consultancy (e.g., Illouz 2007). These industries are constantly producing (new) emotion-regulatory goals suggesting how to use one's emotions in the duty of personal and social success. Some have argued that emotions are also increasingly traded and commodified as market-like goods and services, as well as that cultural practices produce emotional needs that may be satisfied by purchasing certain goods and services (Ehrenberg 2010; Illouz 2007; Neckel 2005). In this way, emotions are socially structured through patterns of emotion regulation and emotion-regulatory goals that are brought about by the dissemination and legitimation of emotion norms and cultural values.

In this vein, knowledge about emotions and their individual and social function has been referred to as a form of capital (Reay 2000). "Emotional capital" is deemed important not only in personal relationships but also in work and organizational contexts as a much-sought interpersonal skill. Presumably, emotional capital, like social, cultural, or economic capital, is unequally distributed in society (Bourdieu 1984) and may depend on age, gender, income, or education. For example, members of certain social units are likely to be better at regulating their emotions than others (Thoits 2004). Understanding how to deal with one's emotions (and those of others) thus is not only a comparative advantage for organizations, but also promotes social distinctions and the reproduction of social inequality:

> emotions are indicators of relative standing; high-status individuals receive respect and liking, low-status persons are offered contempt or disdain. Individuals deliberately manipulate other people's emotions in order to sustain, usurp, upset, or withhold social placement from some and to convey it to others (or themselves). In short, micropolitical emotional exchanges and manipulations are crucial aspects of the creation and perpetuation of social inequality, and the success of these acts depends upon individual's relative possession of the requisite emotional capital.
>
> (Thoits 2004, p. 371)

We can therefore assume that social units endow actors with emotional capital (in particular socially distributed knowledge about important emotion norms) in a way that enables them to align their emotions with the prevailing emotion norms and, where applicable, contribute to the overall calibration and structuration of emotions within a social unit.

Social norms and affective action

Having outlined the interplay of emotions and rational as well as normative action in the previous chapter, and the role of social norms in shaping emotion in the preceding section, this section aims at shedding light on the impact of emotions on the maintenance and enforcement of social norms more generally. As far as questions on the emergence and reproduction of social order are concerned, social norms are important primarily in view of their role in guiding social interactions. Consequently, the arguments outlined in this section rest on the assumption that micro-social orders eventually develop into larger-scale social structures. A number of social norms explicitly relate to problems that frequently occur in social interaction and exchange and, given that they are followed, contribute to establishing stable and long-term frameworks for interactions, for example norms relating to fairness, cooperation, and reciprocity.

When it comes to explaining social norms, a number of questions have until very recently remained unanswered in the social sciences that are of particular interest given the importance of norms outlined in the previous sections. Why do actors follow and comply with social norms at all? And what are the mechanisms on which norm-compliance is based?

> Although no other concept is invoked more frequently in the social
> sciences, we still know little about how social norms are formed, the forces
> determining their content, and the cognitive and emotional requirements
> that enable a species to establish and enforce social norms.
>
> (Fehr & Fischbacher 2004a, p. 185)

Of course, answering these questions in their entirety is well beyond the scope
of this study. However, they shall be examined from the perspective on
emotion and their social structuration outlined in the previous chapters to better
understand how social norms work and how they contribute to the emergence
of social order.

Definitions of social norms tend to be based on somewhat loose assertions.
They are said to be "behavioral rules that are backed by sanctions" (Bendor
& Swistak 2001, p. 1494; Homans 1950), indicators "that something ought or
ought not to be the case" under specific circumstances (Opp 2002, p. 132)
or "customary rules of behavior that coordinate our interactions with others"
(Young 2008). Castelfranchi emphasizes explicit knowledge about norms
and advances the view that "[n]ormative behaviour has to be intentional and
conscious: it has to be based on knowledge of the norm (prescription), but this
does not necessarily imply consciousness and intentionality relative to all the
functions of the norm" (Castelfranchi 2001, p. 31; italics omitted). This means
that propositional representations of behavioral expectations are a critical
component of a social norm and that normative action needs to make conscious
and intentional reference to such a representation.

From a behavioral view, norms have been defined as "patterns of reactions
to behaviors" and exist when actors react differently to some kind of behavior
that causes externalities than to a behavior that causes no such external effects
(Horne & Cutlip 2002, p. 286). Axelrod's standpoint is quite similar: "A Norm
exists in a given social setting to the extent that individuals usually act in a
certain way and are often punished when seen not to be acting this way"
(Axelrod 1986, p. 1097). These definitions do not include the explicit cognitive
representation of a norm. In this vein, some social theorists have compared
normative behavior as a crucial factor in the formation and reproduction of
social order to the internalization of language, which is not chosen in any
conscious, intentional, or even rational way, but is acquired in social contexts
and then in turn determines thoughts and actions. "Like the rules of syntax
identified in transformational grammar, the rules of conduct sought after in
some micro-sociologies are analogous to a level of deep structure of human
behaviour, acquired by the individual through socialization" (Knorr-Cetina
1981, p. 4; see also Callero 1991, p. 52; Giddens 1984, p. 170).

Elster, for example, distinguishes social from moral, legal, and personal
norms not based on their prescriptive qualities, but rather in view of their
enforcement, social distribution, and emotional connotation. According to him,
key characteristics of social norms are that they are outcome-oriented, socially
distributed, and enforced through sanctions (Elster 1999, p. 145 f.). Sanctioning

of deviant behavior is indeed a central element found in many concepts of social norms. Although not all authors include (positive or negative) sanctions in their definitions (Opp 2002), there is broad consensus on the importance of sanctions as an essential element in explaining conformity.

Questions concerning the representation of social norms remain debated in sociology. Coleman takes the view that norms are characteristics of a social system and not of the actors constituting such a system (Coleman 1990). As characteristics of a macro system, norms constrain courses of action and reduce them to a set of socially accepted options. Despite their "location" at the social macro level, they are nevertheless based on actors' goal-oriented rational actions or emerge as a result of those actions (Coleman 1990). Others have taken the view that norms are "mental objects" based on propositional attitudes and can be defined as configurations of beliefs and goals. This, however, leaves outstanding the crucial question of how normative beliefs are turned into normative goals that inform (norm compliant) action (Conte & Castelfranchi 1995, p. 192).

This "foundational theoretical problem of the social sciences" (Castelfranchi 2001, p. 6) is frequently approached by looking at the emergence and functions of norms. One assumption is that norms are functional both in individual and social collective terms. Norms as elements of a macro-level system encourage certain behaviors and sanction others. Given that norms emerge within specific social units, for example through intentional norm setting or as unintended consequences of action, it would be somewhat implausible to assume that they are—by and large—disadvantageous for the functioning of the social unit. Thus, even the unintended emergence of norms, such as through regular patterns of successful interaction and exchange or processes of habituation, internalization, and institutionalization, include functional components that are conducive to the establishment and reproduction of social order (e.g., Berger & Luckmann 1969). Aside from these social functions of norms, some scholars also assume a number of individual functions. Norms relieve actors of the burden of certain decision-making problems and reduce contingencies in social interaction. Norms make actions of others more predictable and in general reduce uncertainties in interactions. Actions of others are simply more predictable in situations governed by norms than in other cases (Popitz 1980).

Taking these functional explanations for the existence of and compliance with norms, it is increasingly popular in sociology to conceive of norm compliance in terms of individually or collectively rational behavior. No doubt, it seems rational to reduce complexities and uncertainties in interactions and to avoid sanctions in cases of non-compliance. Some have therefore argued that norms can be explained from a rational choice perspective and exist because they increase the efficiency of coordination and the overall welfare of the norm beneficiaries (Coleman 1990; Ellickson 1991). According to this view, norms exist because they are rational solutions to problems of cooperation and coordination in social interaction (or are the institutionalized results of such solutions). If a sufficiently large number of actors subscribe to these norms, they are supposed to promote (long-term) increases in collective utility. Rational theories of social norms thus also explain the function of sanctions: Deviant behavior decreases the number of actors that contribute to some public good and thus attenuates the collective utility function. Many of these features of norms have been corroborated by game theory inspired empirical research (see, e.g., Hechter & Opp 2001).

However, rational accounts of norm compliance are susceptible to the same criticism as rational choice theory more generally. Elster argues that the rational model is suited to explaining only a narrowly defined part of normative behavior and rejects the notion that social norms are "'nothing but' instruments of individual, collective or genetic optimization" (Elster 1989, p. 102). According to rational choice accounts, norms are outcomes of rational calculus. Because of their conditionality, they are employed strategically and declared valid or invalid at will. They are merely the means that cover up the goals of maximizing individual utility, for example fairness norms in social exchange dilemmas. However, in such situations, it is crucial that all actors simultaneously declare the fairness norm valid and forego individual utility maximization (Elster 1989), an assumption that is hard to justify based on rational calculus alone.

Furthermore, the view that norm compliance is brought about by rational deliberations to avoid sanctions, is hardly tenable either. Social norms are also followed when there is no possibility of external sanctions (or reputation building). Importantly, once norms have been internalized, they are followed even when non-compliance cannot be sanctioned:

> I don't pick my nose when I can be observed by people on a train passing by, even if I am confident that they are all perfect strangers whom I shall never see again and who have no power to impose sanctions on me. I don't throw litter in the park, even when there is nobody around to observe me. If punishment was merely the price tag attached to crime, nobody would feel shame when caught. People have an internal gyroscope that keeps them adhering steadily to norms, independently of the current reactions of others.
>
> (Elster 1989, p. 104 f.)

Another problem with the (rational) role of sanctions arises when asking the question why actors sanction deviant behavior in the first place, even when deterrence of future harm is not possible or when third parties carry out the sanction (sometimes referred to as "altruistic punishment," Fehr & Gächter 2002). One answer is frequently seen in the concept of "meta norms" that guide sanctioning in cases of deviant behavior (Axelrod 1986). From a rational choice perspective, it would be advantageous to sanction deviant actors as long as the costs of sanctioning remain lower than those that would be incurred by being sanctioned oneself. However, it is scarcely plausible to assume that sanctions based on meta norms will still be taking place even after a second or third iteration—nobody takes issue with an individual who has failed to sanction someone who has failed to sanction someone who in turn has failed to sanction the actual norm violator. Consequently, there have to be other mechanisms that motivate actors to sanction and punish others (Elster 1989, p. 104).

A further objection to rational choice explanations of social norms relates to their contribution to individual utility. There are undoubtedly a number of norms that facilitate increases in individual utility, for example because they reduce complexity and contingency and facilitate decision-making in certain situations. However, referring to threats, deterrence, and retaliation, it can be shown that compliance with social norms cannot solely be explained on the basis of individual utility. For example, given that others can assume that, if affronted or treated unfairly, I will always seek revenge *regardless of the costs* involved in taking revenge, they will carefully consider any affront or unfair treatment. If, on the other hand, others can assume that I will only retaliate when the costs of retaliation are not too high and it is rational and utility maximizing to retaliate, then retributive action will seem unlikely, in particular when the costs are high (Elster 1989). Other examples of "irrational" normative behavior are individual contributions to public goods or the voting paradox. These examples confront rational choice accounts with the problem that actors cannot voluntarily and freely decide to behave irrationally. In this vein, Elster (2004)—paraphrasing Max Weber—notes that "a social norm is not like a taxi from which one can disembark at will" (p. 43). In social reality, actors follow social norms even if they do not contribute to increases in utility.

Other arguments articulated by rational choice proponents relate to the collective rationality of norms. As already indicated, social norms in many cases are supposed to be socially functional and structure reinforcing in nature. In addition, some advocates even assume that social norms contribute to Pareto optimality. From this point of view, no actor in a society characterized by a specific social norm is worse off and a significant number of actors are better off than in a society in which this specific norm does not exist. Norms are therefore frequently seen as social institutions whose primary purpose is to compensate for market failure (Elster 1989, p. 107 f.). However, a large number of social norms do not fulfill these conditions. Rules of etiquette or fashion norms, for example, are certainly not Pareto optimal. They may indeed be socially functional as means of social distinction, but collective rationality is hardly one of their characteristics. Norms against certain types of interpersonal relations do not meet the conditions of collective rationality either. Those against homosexuality are not only increasingly losing their validity, but also have no rational characteristics of any kind whatsoever. The use of money is also strongly regulated by social norms. Elster (1989) gives the example of queuing. Usually, one cannot purchase one's position in a queue by offering money to a person ahead of oneself, even though this might constitute a Pareto improvement in many cases.

In light of these objections to purely rational accounts of social norms, some have suggested to relate social norms to other explanatory principles outside of rational deliberation or utility maximization, for instance certain psychological or bodily dispositions or universals (Elster 1989, p. 115; Popitz 1980). One such possibility is to account for the role of emotions in norm compliance. This is by no means a new strategy in explaining social norms, as some philosophical and psychological explorations have shown (e.g., Haidt & Joseph 2004; Nichols 2004; Prinz 2007). In fact, Elster (1996, p. 1389) includes them in his definition of social norms as "sustained by internalised emotions." Nevertheless, given the way emotions have been conceptualized in the preceding chapters, they might offer a deeper understanding of social norms as critical facilitators in the reproduction of social order.

In most approaches to explaining norm compliance, sanctions play an important role or are even key elements of definitions of norms. They contribute to compliance as costs or negative incentives linked to certain behaviors. Thus, non-compliance generates costs not only for deviant actors, but also for those who punish norm violators. This applies in particular to third parties who sanction violators even though they are not immediately affected by the norm infringement and do not directly benefit from sanctioning. Thus altruistic punishment and compliance with norms in the absence of material sanctions are cases in which rational calculation can for the most part be eradicated from the equation. They are therefore the most interesting ones when looking for the emotional basis of norm compliance.

Elster (2004) has suggested that sanctions in general—from expressions of disapproval via social exclusion to the refusal of resources and cooperation—affect norm violators not merely because of the material consequences but also because they act as vehicles for the expression of negative emotions such as contempt, anger, irritation, disgust, or disdain. Becoming the object of such negative emotions can be experienced as significantly more far-reaching than the loss of purely material resources, since negative emotions indicate the loss of significant social resources, such as social support and the willingness to cooperate. This view is further supported by the fact that norm violators generally react to these sanctions, as well as to their very own norm-violating behavior, with negative emotions, in particular shame and guilt.

Shame is particularly important here because it also reflects the sanctioning party's point of view and involves an actor's social self in the interpretation of a sanction (cf. Ketelaar & Au 2003). As the "master emotion of everyday life" (Scheff & Retzinger 2000), shame plays a key role in sanctioning since it indicates threats to social bonds and gives rise to fear of the disruption of social bonds (Scheff 1997). "The point that shame is a response to bond threat cannot be emphasized too strongly, since in psychology and psychoanalysis there is a tendency to individualize shame, taking it out of its social matrix" (Scheff & Retzinger 2000). This function of shame as a sort of "moral gyroscope" (Scheff & Retzinger 2000) is also critical because it signals norm violations in a non-conceptual "pre-linguistic" and almost "bodily" way. This position is further reinforced by studies showing that the "social pain" experienced in episodes of shame, guilt, and social exclusion recruits the same neural circuits that are involved in processing physical pain (Eisenberger *et al.* 2003; Eisenberger & Lieberman 2004). Although these findings do not reflect reactions to a norm violation par excellence, they do reinforce the role of shame, guilt, and exclusion as effective negative incentives whose aversive potential is obviously comparable to that of physical pain. They further reveal the influence that shame, guilt, and exclusion (or the expectation thereof) can have on the courses of social action. If certain actions, such as norm violations for example, are highly likely to give rise to such emotions, actors are most likely highly motivated to refrain from these actions. Even if it would seem, at a first glance, that these negative emotions affect the evaluation of options for action and decision-making in the same way as costs, calculations of this kind cannot be conceived of as purely rational, because the effects of emotions are mostly non-propositional and non-conceptual and are to a certain extent beyond voluntary control.

Against this background, sanctions can also be seen in a different light. According to Elster (1999), the decisive factor determining the effectiveness of sanctions is not the costs incurred by the norm violator (i.e., the "price" an actor is willing to pay for a norm violation), but rather the costs a punisher is willing to take to implement a sanction. To punish a norm violator, actors are willing to accept substantial costs that often by far exceed the damage done by a norm violator. Given that both material and social sanctions are vehicles for negative emotions towards the norm violator, then those emotions will most probably be expressed more strongly the higher the costs for the sanctioning party are. Accordingly, the shame felt by a norm violator is also a function of the costs incurred by the punisher and both indicate the severity of a transgression (Elster 1999, 2004).

In sum, there are a number of good arguments in support of the notion that norm compliance is closely linked to emotions and sanctions, for both norm violators and the sanctioning party. Aside from their unpleasant subjective experience, shame and guilt serve as internalized indicators of norm transgressions. Anger, disdain, contempt, and other negative emotions felt by those affected by a norm violation are—in addition to their unpleasant subjective feeling—additional triggers and reinforcers of shame and guilt felt by violators. These emotions thus are strong negative incentives to comply with social norms.

They also explain why social norms are followed even in the absence of other actors and without the possibility of external sanctions. In these cases, external sanctions have been closely linked to negative emotions: External sanctions and constraints have been internalized and become self-constraints. This argument is supported by several studies, some of them combining game theory and social exchange with neuroimaging methods to uncover the emotional dimension of norms. These studies take advantage of the fact that in social exchange, cooperation is usually based on a norm of conditional cooperation. This norm stipulates cooperation only on the condition that the other player also cooperates. This norm mainly results in stable exchange strategies such as tit-for-tat, although there are always free-riders who never cooperate (Axelrod & Hamilton 1981; Fehr & Fischbacher 2004a).

Standard explanations for these strategies include, among others, direct and indirect reciprocity and "costly signaling" (Axelrod & Hamilton 1981). However, it is still largely unknown why actors comply with norms and cooperate even in anonymous one-shot interactions with no possible effects for reputation or any other future benefits (Fehr & Gächter 2002). Research has shown that the willingness to cooperate increases when third-party punishment is possible (Bendor & Swistak 2001; Fehr & Fischbacher 2004a; Fehr & Fischbacher 2004b; Fehr & Gächter 2002). One way to illustrate the impact of third-party sanctioning (or "altruistic punishment") is an extended version of the prisoner's dilemma in which a third player, who only observes the actions of the other players, can punish non-cooperative behavior. Sanctions reduce the gains of the non-cooperative, norm-violating player and are costly for the punisher. All players are anonymous and do not engage in any further interactions with each other. Results show that about 60 percent of observers punish non-cooperative behavior, even though they do not obtain any "rational" advantage from it and indeed incur costs (Fehr & Fischbacher 2004b).

Another way to illustrate the importance of third-party punishment for norm compliance is the Ultimatum Game (Güth *et al.* 1982). In this game, two players have to divide an amount of money given to one of the players (the proposer). The proposer has to make an offer to the second player (the responder) who can either accept or decline the offer. If the responder accepts the offer, the money is shared in the proposed way. If, however, the responder declines, neither of the players receives any money. Proposers therefore want to make offers placing them in a good position but not causing the responder to decline. Usually, about two thirds of proposals are between 40 and 50 percent of the total sum, while only 4 percent of players offer less than 20 percent of that sum. More than half of all participants reject offers of less than 20 percent. According to the rational choice model, however, responders would have to accept any offer (Sanfey *et al.* 2003).

Apart from the fact that Ultimatum Game bargaining and the prisoner's dilemma (with third-party punishment) clearly demonstrate the limits of rational choice models, they also provide evidence for the importance of sanctions in norm compliance and enforcement. Rejections of unfair offers and punishment of non-cooperative players can be interpreted as costly sanctions for the violation of fairness norms and the norm of conditional cooperation— without any possibility of deterrence or reputation building. Therefore, the "willingness to sanction norm violations and non-cooperative behavior is crucial for the maintenance of social order. Such sanctions sustain the viability of a myriad of informal agreements in markets, organizations, families, and neighborhoods" (Falk *et al.* 2005, p. 2028).

If norm compliance and enforcement in certain situations can be explained by sanctioning, then the question arises why actors sanction norm violations at all under these circumstances (which is costly, and offers no promise of reputation or future deterrence)? Emotions offer one possible explanation. Studies have shown that non-cooperative behavior and free-riding, i.e. the violation of fairness and cooperation norms, cause cooperative actors to experience strong negative emotions, such as anger, rage, and annoyance, which are also anticipated by free-riders. The negative emotions produced by a norm violation are experienced and expressed all the more strongly the more clearly someone deviates from a social norm (Fehr & Gächter 2002, p. 139). These results are supported by a neuroimaging study showing that offers in the Ultimatum Game that are perceived as unfair are accompanied by increased activity in brain areas involved in the processing of aversive reactions such as pain, hunger, and thirst, and in the appraisal and representation of basic affective reactions, in particular rage, anger, and disgust (Sanfey *et al.* 2003). In this study, proposers in the Ultimatum Game were either human actors or a computer program. Results show that when human players make unfair offers, the affective reactions are significantly more severe than when offers are made by the computer. This difference suggests that the activation of brain areas implicated in processing negative affective reactions is not a result of the monetary loss as such, but rather a consequence of appraising a specific social action as norm violating (Sanfey *et al.* 2003, p. 1756).

Likewise, neuroimaging studies investigating the reactions of punishers support the argument that emotions play a decisive role in the enforcement of social norms. One study looked into the motivations underlying punishment in a modified trust-game (de Quervain *et al.* 2004). Results show that the punishment of norm violations involves increased activity in brain areas implicated in reward processing. This suggests that punishment is associated with gratification and satisfaction and other positive affects which may count as (additional) motivators of sanctioning (see also Rilling *et al.* 2002; Stephen & Pham 2008). De Quervain and colleagues (2004) emphasize that increased activity in these brain areas seems to reflect the *expected* gratification, so that positive affective reactions are presumably the cause rather than the consequence of punishment.

In sum, the role of emotions in norm compliance and enforcement marks a further step in understanding emotions as bi-directional mediators between social action and social structure. Emotions are themselves decisively shaped by social norms, but at the same time contribute to the enforcement of norms. From the perspective of a norm violator, negative emotions such as shame and guilt can be seen as sanctions in much the same way as material sanctions. These intrinsically negative emotions are so effective because their aversive potential is comparable to that of physical pain and actors are therefore strongly motivated to avoid experiencing these emotions. Importantly, emotions contribute to norm compliance even in the absence of other actors and external sanctions. Thus, in the light of their social structuration and calibration outlined in the previous chapters, emotions such as guilt and shame are an important means of primary social control. Moreover, negative emotions such as anger and rage have been shown to promote punishment even by third parties, which is also critical to the enforcement of norms. Finally, punishment in itself is linked to basic positive affects and gratification.

Concluding remarks

This study investigated the role of emotions in the emergence and reproduction of social order. I discussed the main thesis that emotions are a bi-directional mediator between social action and social structure. In an interdisciplinary approach drawing on theory and evidence from disciplines such as social neuroscience and psychology, I have illustrated the social shaping and calibration of emotion, their influence on social action, and their role in social interaction. This interdisciplinary perspective is critical to the overall argument advanced in this study, since it complements a genuinely sociological perspective with in-depth accounts of the biological and psychological components of emotion and their contributions to understanding one of the fundamental problems of the social sciences, namely the micro–macro link and the question of the relationship between self and society.

Ever since its founding days, sociology has been dealing with a dilemma that is inherent to the concepts of social action and social order. On the one hand, sociologists are faced with autonomous agents that act intentionally and (at times) self-interestedly. On the other hand, sociology is confronted with large-scale social structural arrangements and systems of symbolic order whose characteristics cannot be fully reduced to individual minds and actions, but which undoubtedly influence those actions. Hence, the intricate problem is this: If social actors are largely autonomous and self-interested, how do social structures emerge that require regular and interlinked patterns of social action? In response to this problem, sociology has developed an impressive line of theories and arguments, most of which assume some kind of "invisible hand" bringing about robust (intended as well as unintended) social structural effects. To give but two of the most prominent examples, Giddens (1984) acknowledges this problem in his account of the "duality of structure" and Bourdieu (1984) deals with it using the concepts *habitus* and social field.

Looking at this puzzle from the perspective of autonomy and self-interest, we arrive at models highlighting intentional and instrumental rational action in bringing about larger scale social structural effects and social order. However, as illustrated in Chapter 3, this approach is fraught with a number of pitfalls and weaknesses inherent to the rational choice paradigm. Approaching the puzzle by softening the assumption of autonomous action and instead stressing the influence of social order and social structure on behavior, we arrive at accounts of a social action that is mostly constrained by social norms. However, as discussed in Chapter 4, accounts of behavior as constrained by social norms are equally problematic in some respects, especially when looked at from a rational choice perspective, but also when relying on an "oversocialized conception of man" (Wrong 1961).

Among other things, this study has suggested that emotion shed light on these problems when integrated into the various explanatory frameworks related to the *homo economicus* and the *homo sociologicus*. To do this, however, it is essential to develop a perspective on emotion that goes beyond the one frequently adopted in sociology, which is primarily concerned with culture and social constructionism at the level of discourse, conceptual and declarative knowledge. I have argued that, in order to uncover the role of emotions as bi-directional mediators between social action and social structure, we have to take a closer look at the biological and psychological components of emotion and uncover their plasticity and susceptibility to systematic influences of culture and society.

In Chapter 2, I have argued that the social construction of emotions begins at the biological level of the basic affect system that produces what has become known as "core affects" (e.g., Barrett & Bliss-Moreau 2009). I have argued that basic affective reactions are not as "hardwired" and biologically determined as claimed in some neuroscience and social science camps. Admittedly, they are based on a simple and universal "motivational metric" (Cacioppo & Gardner 1999), but, with very few exceptions, the association of some sort of "stimulus" and ensuing affective reaction is not part of a genetically determined behavioral program, but rather the outcome of learning, socialization, and internalization. Even more obviously, the same applies to complex emotions such as shame, pride, and guilt. Although these emotions also rely on core affective reactions, they require conceptual knowledge, which is the outcome of culture and social practices. Importantly, the complexity of certain emotions does not necessarily mean less rapidity, less immediacy, and less embodied grounding in their processing, nor does it imply as much for the associated action tendencies. Even complex emotions can be triggered schematically and automatically, much like basic affective reactions and without conscious involvement or awareness.

Sociology can benefit from this perspective because it promotes the synthesis of two still largely opposing paradigms. On the one hand, as Chapter 2 has shown, it accounts for "positivist" approaches to the sociology of emotion by emphasizing the non-conscious, automatic, and biological processes of emotion. On the other hand, it acknowledges social constructionist and cultural views by showing that emotions are socially structured and calibrated at levels frequently dismissed as biologically determined and immune to social influence.

Aside from this social plasticity and openness to social and cultural shaping, the impact of emotions on social action is the second crucial factor in explaining their role in the emergence and reproduction of social order. In contrast to other forms of "conditioned" behavior or socially constrained action, emotions do not simply and directly give rise to a particular course of action, but through their influence on cognitions and bodily reactions generate certain action *tendencies*. As a result, we can avoid an overly deterministic account of social action while still allowing for several principles of the bounded rationality of decision-making. In consequence, I have argued that emotions increase the probability that certain courses of action will be taken more regularly than others and thus contribute to the formation of larger-scale patterns of action and social structure.

In attending to structure–agency links, sociology has frequently and classically—although not exclusively—emphasized purposeful and intentional action as opposed to mere "instinct-driven" behavior. The implication of this position is that much of sociological research has conceived of social action as *choice between alternatives* and focused on the question of why actors behave in one way rather than in another. The answers given to date have concentrated primarily on two aspects. The first is the purpose or the goal that is pursued when a choice is made. An individual can decide, for example, to take a particular action in order to benefit someone else or to protect oneself from harm. The second aspect concerns the rationale on which goal attainment is based. One can decide between alternatives on the basis of, for example, means–ends rationality, moral convictions, normative obligations, or out of habit. Crucially for this view, there has to be a sufficiently large number of actors whose actions are based on these principles and are not self-contradictory.

Both aspects are reflected in the concepts of *homo economicus* and *homo sociologicus*—and both models are beset with the problems I have discussed in Chapter 3. Emotions, I have suggested, help to attenuate some of these problems. Rational and normative action is based on knowledge, opinions, beliefs, attitudes, and goals that are—according to the premises of cognitive sociology and the sociology of knowledge—shaped and structured in various social contexts. Thus, culture and society in some way are always reflected in rational and normative accounts of action. Emotions constitute a mechanism that strengthens these cognitive and social structures (of action) by establishing experience-based bodily links between thought and action. They systematically intervene in the process of choosing between alternatives in such a way that actors may not even be aware of the fact that they *have* alternatives at all. And, importantly, the nature of this influence is shared across many individuals within a social unit. This way, emotions promote something that might well resemble "instinctive" behavior, albeit this could not be more different from any biologically hardwired form of behavior. The concept of *affective action* I have developed in Chapter 3 summarizes the role of socially structured emotions in social action and decision-making and highlights their repercussions for sociality. The concept is not only useful to more traditional accounts of structure–agency and micro–macro problems, but may also inform more recent approaches to explaining regular patterns of action rooted in cultural theory, in particular theories of social practices and praxeological accounts of action (e.g., Bourdieu 1998; Reckwitz 2002; Schatzki 1996; Turner 1994).

Attempts at uncovering the consequences of socially structured emotions for social units and social order solely on the grounds of affective action would be insufficient regarding the role of emotion in social encounters and face-to-face interactions. This importance becomes particularly manifest in the involuntary nature of emotional facial expressions, the rapid and automatic decoding and recognition of emotion expressions, and the supposition that this decoding is more effective when sender and receiver have comparable social and cultural backgrounds. Moreover, it is evident in emotional contagion, i.e. the automatic and effortless interindividual transfer of affect and emotion.

These mechanisms make the social structuration of emotion publicly available and accessible to other actors and contribute to its interindividual dissemination. Just as emotions can be expressed verbally in different dialects and accents, they are expressed in "facial dialects" that are specific to social units. In other words, emotion expressions are "calibrated" to the social environments in which actors are socialized. The public nature of such expressions enables other actors to draw conclusions not only about the underlying emotions, but also about the appraisals and cognitive evaluations of events. Emotions therefore critically contribute to the establishment of intersubjectivity and mutual understanding and in fact constitute what might be called *interaffectivity*, i.e. the affect-based implicit and "empathic" knowledge of others. Moreover, emotional contagion contributes to the alignment of emotions across actors. This emotional convergence also promotes the emergence of collective emotions across larger numbers of individuals (von Scheve & Ismer 2013).

Aside from this automatic and mostly pre-conscious social calibration of emotions, I have emphasized that they are also shaped and constrained by social norms. Emotion norms prescribe which particular emotions should be felt and expressed in specific situations. Importantly, socially structured emotions can be conceived of as "descriptive norms," i.e. as perceptions of what individuals "usually" or "normally" feel in a specific situation. In contrast, prescriptive (injunctive) emotion norms directly refer to (socially structured) emotions and imbue them with a normative status and specific behavioral expectations that can be enforced by sanctions, much in the same way as with other norms and behaviors. This social control of emotions is met with various forms of emotion regulation and emotion work motivated by emotion norms as regulatory goals.

Moreover, I have illustrated that social norms do not only influence emotions, but, conversely, emotions also promote the enforcement and maintenance of social norms. As internalized (and mostly negative) sanctions, emotions foster norm compliance even in the absence of external material sanctions. Emotions such as guilt and shame are often considerably more effective as sanctions than other forms of punishment. As positive incentives, emotions also contribute to the implementation of third-party sanctions or altruistic punishment, thereby reducing the likelihood of emotional deviance.

This account of emotion in social interaction extends and enriches sociological theory, in particular regarding symbolic interactionism, which can be complemented with a *sociologically meaningful* theory of "non-symbolic interaction" (Blumer 1969). This concerns, on the one hand, symbolic inter-actionist theories of emotion, which primarily focus on the staging and dramaturgical presentation of emotion in social encounters. The breakup of the dichotomy between universal and socially constructed components of facial expression does suggest that expressions, although they occur rapidly, involuntarily, and automatically, and are usually "understood" without much interpretative effort, are yet closely associated with actors' social and cultural background and thus socially meaningful. On the other hand, interaffectivity can well be understood as a component of processes of objectivation and institutionalization. The involuntary expression, decoding, and transmission of socially structured emotions and the manifold influences emotions exert on knowledge and cognition can be seen as constituting (part of) the non-conscious bases of the processes outlined by Berger and Luckmann (1969).

One of the challenges for future studies is indeed the empirical investigation of the theoretical propositions developed in this work. Here, the objective has to be to further specify the understanding and possible operationalization of social structure and systems of social order and how they manifest in society. Here, it seems possible to employ traditional indicators of social differentiation, social inequality, and stratification, but also more horizontal measures such as functional or institutional arenas, lifestyles, social milieus, or ethnic, national, and religious groups.

Some empirical studies have indeed attended to such indicators. For example, a number of studies looked at associations between socio-economic status and certain emotions, in particular anger (Collett & Lizardo 2010; Marby 1999; Rackow *et al.* 2012; Schieman 2003). Also, research in medical sociology and health psychology has attended to the question whether (negative) emotions are systematically implicated in the negative relation-ship between socio-economic status, social inequality and health (Adler *et al.* 1994; Marmot 2004; Wilkinson & Pickett 2009). Here, emotions such as shame, anger, and anxiety are suspected to be possible mediators (Gallo & Matthews 2003; Roy 2004; Wilkinson 1999). Some have even argued that modern societies lead to the "chronic" experience of these emotions and in some cases and can lead to illnesses such as depression and chronic fatigue (Ehrenberg 2010). Still other research indicates a negative association between socio-economic status and the development of brain areas that subserve important cognitive and probably also affective functions (Hackman & Farah 2009; Hackman *et al.* 2010).

In sum, the implications of the arguments outlined in this study for the general understanding of the emergence and reproduction of social order, which are primarily grounded in insights on the nature, social construction, and functions of emotions, are fundamentally rooted in the interdisciplinary nature of the investigation. In the past, attempts at incorporating findings from biology and psychology into sociological explanations were often met with hesitation or outright denial. In the chapters of the present book, I hope I was able to attenuate such concerns and to show that looking at the embodied aspects of mind, feeling, and thinking can be made fruitful for sociology.

Notes

Introduction

1. Terminological note: I will use the term "social structure" primarily to refer to various constellations of actors, such as certain patterns of interaction or systems of stratification, as expressed, for example, in allocations of status and power; "social order" first and foremost denotes the orders of sense-making and meaning-making and the underlying "cognitive structures," as expressed in certain norms and values. The terms "social units" or "social entities" are used to refer to aggregates of actors that can be characterized by social structures and social orders.

1. Self, society, and emotion

1. But see Thoits (2007) for a sociological critique.

2. Socially structured emotions

1. James (1884) in fact proposed that the physiological reaction *is* the emotion. Here, it becomes difficult to conceptually distinguish between questions of what an emotion is (as discussed in the previous chapter) and how emotions are generated. Nevertheless, the focus in this chapter will be on the generation or elicitation of emotion and their biological bases in relation to society and culture.
2. Some authors consider the concept of the "limbic system" to be problematic since it consists of a number of sub-systems without clear functional, anatomical, or psychological defining criteria (LeDoux 2000). Others take the view that, if used appropriately, the term is useful to distinguish cortical and subcortical emotion processing (Panksepp 2003a).
3. See Cacioppo and Berntson (1999), Cacioppo, Frankel, and Camras (2004), Davidson and Irwin (1999), LeDoux (1996), Panksepp (1998), and Rolls (1999).
4. See Barrett, Mesquita, Ochsner, and Gross (2007), Britton and colleagues (2005), Damasio (1994), Davidson (2004), Kringelbach and Rolls (2004), Ochsner and Barrett (2001), and Rolls (2004).
5. See Barrett, Mesquita and colleagues (2007), Berridge and Winkielman (2003), Cosmides and Tooby (2000), Damasio and colleagues (2000), Darwin (1872), Kober and colleagues (2008), LeDoux (2000), Panksepp (1998), Phelps and LeDoux (2005), Turner (2000).
6. See Davidson (2003a, p. 129), Duncan and Barrett (2007), Eder, Hommel, and Houwer (2007), Gray (1990), LeDoux (1993, 2000), Parrott and Schulkin (1993), Rolls (1990, 1999), Storbeck and Clore (2007).
7. However, some appraisal theorists have suggested that appraisals are also a component or even a consequence of emotion (see Roseman & Smith 2001).

8. Some of these theories have been referred to as *goal-relevance theories* (Clore *et al.* 1994, p. 332).
9. Similar concepts are *background emotions* (Barbalet 1998, p. 29) and *dispositional emotions* (Elster 1999, p. 244).

4. The affective structure of social interaction

1. This section is based on von Scheve (2012b).
2. This section is in part based on von Scheve (2012a).

References

Adler, N. E., Boyce, T., Chesney, M. A., Cohen, S., Folkman, S., Kahn, R. L., Syme, S. L. (1994). Socioeconomic status and health. The challenge of the gradient. *American Psychologist, 49(1)*, 15–24.

Adolphs, R. (2002a). Recognizing emotion from facial expressions: Psychological and neurological mechanisms. *Behavioral and Cognitive Neuroscience Reviews, 1(1)*, 21–61.

Adolphs, R. (2002b). Neural systems for recognizing emotion. *Current Opinion in Neurobiology, 12(2)*, 169–177.

Adolphs, R., Damasio, A. R. (2000). Neurobiology of emotion at a systems level. In Borod, J. C. (Ed.), *The neuropsychology of emotion*. New York: Oxford University Press, 194–213.

Alexander, J. C., Giesen, B. (1987). From reduction to linkage: The long view of the micro–macro debate. In Alexander, J. C., Giesen, B., Münch, R., N. J. Smelser (Eds.), *The micro–macro link*. Berkeley, CA: University of California Press, 1–42.

Anderson, A. K., Phelps, E. A. (2001). Lesions of the human amygdala impair enhanced perception of emotionally salient events. *Nature, 411*, 305–309.

Anderson, J. R., Bower, G. H. (1973). *Human associative memory*. Washington, DC: Winston.

Archer, M. S., Tritter. J. Q. (2000). *Rational choice theory. Resisting colonization*. London: Routledge.

Armon-Jones, C. (1986). The thesis of constructionism. In Harré, R. (Ed.), *The social construction of emotions*. Oxford: Blackwell, 57–82.

Arnold, M. B. (1960). *Emotion and personality*. Vol. 1, 2. New York: Columbia University Press.

Augoustinos, M., Walker, I. (1995). *Social cognition*. London: Sage.

Averill, J. R. (1980). A constructivist view of emotion. In Plutchik, R., Kellerman, H. (Eds.), *Emotion: Theory, research, and experience*. Vol. 1, *Theories of emotion*. New York: Academic Press, 305–339.

Axelrod, R. (1986). An evolutionary approach to norms. *American Political Science Review, 80(4)*, 1095–1111.

Axelrod, R., Hamilton, W. D. (1981). The evolution of cooperation. *Science, 211*, 1390–1396.

Barbalet, J. M. (1992). A macro sociology of emotion: Class resentment. *Sociological Theory, 10(2)*, 150–163.

Barbalet, J. M. (1998). *Emotion, social theory, and social structure*. Cambridge: Cambridge University Press.

Barbalet, J. M. (2002). Introduction: Why emotions are crucial. In Barbalet, J. M. (Ed.), *Emotions and sociology*. Oxford: Blackwell, 1–9.

Bargh, J. A. (1997). The automaticity of everyday life. In Wyer, R. S. (Ed.), *The automaticity of everyday life*. Mahwah, NJ: Erlbaum, 1–61.

Bargh, J. A., Chartrand, T. L. (1999). The unbearable automaticity of being. *American Psychologist, 54(7)*, 462–479.

Bargh, J. A., Ferguson, M. L. (2000). Beyond behaviorism: On the automaticity of higher mental processes. *Psychological Bulletin, 126(6)*, 925–945.

Barnard, P. J., Teasdale, J. D. (1991). Interacting cognitive subsystems: a systemic approach to cognitive-affective interaction and change. *Cognition and Emotion, 5(1)*, 1–39.

Barrett, L. F. (2006). Are emotions natural kinds? *Perspectives on Psychological Science, 1(1)*, 28–58.

Barrett, L. F. (2012). Emotions are real. *Emotion, 12(3)*, 413–429.

Barrett, L. F., Bliss-Moreau, E. (2009). Affect as a psychological primitive. In Zanna, M. P. (Ed.), *Advances in Experimental Social Psychology*. Vol. 41. Burlington, VT: Academic Press, 167–218.

Barrett, L. F., Wager, T. D. (2006). The structure of emotion. Evidence from neuroimaging studies. *Current Directions in Psychological Science, 15(2)*, 79–83.

Barrett, L. F., Bliss-Moreau, E., Duncan, S. L., Rauch, S. L., Wright, C. I. (2007). The amygdala and the experience of affect. *Social Cognitive and Affective Neuroscience, 2(2)*, 73–83.

Barrett, L. F., Mesquita, B., Ochsner, K. N., Gross, J. J. (2007). The experience of emotion. *Annual Review of Psychology, 58*, 373–403.

Barrett, L. F., Ochsner, K. N., Gross, J. J. (2007). On the automaticity of emotion. In Bargh, J. (Ed.), *Social psychology and the unconscious: The automaticity of higher mental processes*. New York: Psychology Press, 173–217.

Baumeister, R. F., Vohs, K. D., DeWall, N. C., Zhang, L. (2007). How emotion shapes behavior: Feedback, anticipation, and reflection, rather than direct causation. *Personality and Social Psychology Review, 11*, 167–203.

Bayley, P. J., Frascino, J. C., Squire, L. R. (2003). Robust habit learning in the absence of awareness and independent of the medial temporal lobe. *Nature, 436*, 550–553.

Bechara, A. (2004). The role of emotion in decision-making: Evidence from neurological patients with orbitofrontal damage. *Brain and Cognition, 55(1)*, 30–40.

Bechara, A., Damasio, H., Damasio, A. R. (2000). Emotion, decision-making and the OFC. *Cerebral Cortex, 10(3)*, 295–307.

Bendor, J., Swistak, P. (2001). The evolution of norms. *American Journal of Sociology, 106(6)*, 1493–1545.

Benton, T. (1991). Biology and social science: Why the return of the repressed should be given a (cautious) welcome. *Sociology, 25(1)*, 1–29.

Berger, P. L., Luckmann, T. (1966). *The social construction of reality*. Garden City, NY: Anchor.

Berkowitz, L. (2000). *Causes and consequences of feelings*. New York: Cambridge University Press.

Berridge, K. C., Winkielman, P. (2003). What is an unconscious emotion? (The case of unconscious "liking"). *Cognition and Emotion, 17(2)*, 181–211.

Biccheri, C. (2005). *The grammar of society*. New York: Cambridge University Press.

Biddle, B. J. (1986). Recent developments in role theory. *Annual Review of Sociology, 12*, 67–92.

Bless, H. (2000). The interplay of affect and cognition: The mediating role of general knowledge structures. In Forgas, J. P. (Ed.), *Feeling and thinking. The role of affect in social cognition*. New York: Cambridge University Press, 201–222.

Bless, H. (2001). The relation between mood and the use of general knowledge structures. In Martin, L. L., Clore, G. L. (Eds.), *Theories of mood and cognition*. Mahwah, NJ: Erlbaum, 9–29.

Bless, H., Clore, G., Schwarz, N., Golisano, V., Rabe, C., Wölk, M. (1996). Mood and the use of scripts: Does happy mood make people really mindless? *Journal of Personality and Social Psychology, 71(4)*, 665–679.

Bless, H., Fiedler, K., Strack, F. (2004). *Social cognition. How individuals construct social reality*. Hove: Psychology Press.

Blumer, H. (1969). *Symbolic interactionism*. Englewood Cliffs, NJ: Prentice Hall.

Bock, J., Helmeke, C., Ovtscharoff, W., Gruß, M., K. Braun (2003). Frühkindliche emotionale Erfahrungen beeinflussen die funktionelle Entwicklung des Gehirns. *Neuroforum, 2(3)*, 51–55.

Bourdieu, P. (1984). *Distinction: A social critique of the judgement of taste*. Cambridge, MA: Harvard University Press.

Bourdieu, P. (1992). *The logic of practice*. Stanford, CA: Stanford University Press.

Bourdieu, P. (1998). *Practical reason: On the theory of action*. Stanford, CA: Stanford University Press.

Bourgeois, P., Hess, U. (2008). The impact of social context on mimicry. *Biological Psychology, 77(3)*, 343–352.

Bower, G. H. (1981). Mood and memory. *American Psychologist, 36(2)*, 129–148.

Braun, K. (2011). The prefrontal-limbic system: Development, neuroanatomy, function, and implications for socioemotional development. *Clinical Perinatology, 38*, 685–702.

Brief, A.P., Weiss, H. M. (2002). Organizational behavior: Affect in the workplace. *Annual Review of Psychology, 53*, 279–307.

Britton, J. C., Phan, K. L., Taylor, S. F., Welsh, R. C., Berridge, K. C., Liberzon, I. (2005). Neural correlates of social and nonsocial emotions: An fMRI study. *NeuroImage, 31(1)*, 397–409.

Brothers, L. (1997). *Friday's footprint*. New York: Oxford University Press.

Bruce, V., Young, A. (1986). Understanding face recognition. *British Journal of Psychology, 77(3)*, 305–327.

Buck, R. W. (1984). *The communication of emotion*. New York: Guilford.

Buck, R., Powers, S. R. (2011). Emotion, media, and the global village. In Döveling, K., von Scheve, C., Konjin, E. (Eds.), *The Routledge handbook of emotions and mass media*. London: Routledge, 181–194.

Cacioppo, J. T., Berntson, G. G. (1999). The affect system: Architecture and operating characteristics. *Current Directions in Psychological Science, 8(5)*, 133–137.

Cacioppo, J. T., Gardner, W. L. (1999). Emotion. *Annual Review of Psychology, 50*, 191–214.

Cacioppo, J. T., Berntson, G. G., Larsen, J. T., Poehlmann, K. M., Ito, T. A. (2000). The psychophysiology of emotion. In Lewis, R., Haviland-Jones, J. M. (Eds.), *Handbook of emotions*, 2nd ed. New York: Guilford, 173–191.

Cacioppo, J. T., Larsen, J. T., Smith, N. K., Berntson, G. G. (2004). The affect system. What lurks below the surface of feelings? In Manstead, A. S., Frijda, N. H., Fischer, A. (Eds.), *Feelings and emotions*. New York: Oxford University Press, 223–242.

Callero, P. L. (1991). Toward a sociology of cognition. In Howard, J. A., Callero, P. L. (Eds.), *The self–society dynamic*. New York: Cambridge University Press, 43–54.

Camerer, C. F., Fehr, E. (2006). When does "economic man" dominate social behavior? *Science, 311*, 47–52.

Campos, J. J., Frankel, C. B., Camras, L. (2004). On the nature of emotion regulation. *Child Development, 75(2)*, 377–394.

Cancian, F. M., Gordon, S. L. (1988). Changing emotion norms in marriage: Love and anger in U.S. women's magazines since 1900. *Gender and Society, 2(3)*, 308–342.

Cannon, W. (1927). The James–Lange theory of emotion: A critical examination and an alternative theory. *American Journal of Psychology, 39*, 106–124.

Cannon-Bowers, J. A., Salas, E. (2001). Reflections on shared cognition. *Journal of Organizational Behavior, 22(2)*, 195–202.

Carroll, J. M., Russell, J. A. (1996). Do facial expressions signal specific emotions? Judging emotion from the face in context. *Journal of Personality and Social Psychology, 70(2)*, 205–218.

Castelfranchi, C. (2001). The theory of social functions: Challenges for computational social science and multi-agent learning. *Cognitive Systems Research, 2(1)*, 5–38.

Cerulo, K. A. (2002). Establishing a sociology of culture and cognition. In Cerulo, K. A. (Ed.), *Culture in mind*. New York: Routledge, 1–14.

Chartrand, T. L., Bargh, J. A. (1999). The chameleon effect: The perception–behavior link and social interaction. *Journal of Personality and Social Psychology, 76(6)*, 893–910.

Cialdini, R. (2007). Descriptive social norms as underappreciated sources of social control. *Psychometrika, 72*, 263–268.

Cicchetti, D., Curtis, W. J. (2006). The developing brain and neural plasticity: Implications for normality, psychopathology, and resilience. In Cicchetti, D., Cohen, D. J. (Eds.), *Developmental psychopathology: Developmental neuroscience*. Vol. 2, 2nd ed. New York: Wiley, 1–64.

Coleman, J. S. (1990). *Foundations of social theory*. Cambridge, MA: Harvard University Press.

Collett, J. L., Lizardo, O. (2010). Occupational status and the experience of anger. *Social Forces, 88(5)*, 2079–2104.

Collins, R. (1975). *Conflict sociology*. New York: Academic Press.

Collins, R. (1993). Emotional energy as the common denominator of rational action. *Rationality and Society, 5(2)*, 203–230.

Collins, R. (2004). *Interaction ritual chains*. Princeton, NJ: Princeton University Press.

Conte, R., Castelfranchi, C. (1995). *Norms as mental objects. Proceedings of the 5th European workshop on Modelling Autonomous Agents in a Multi-Agent World*. Heidelberg: Springer, 186–196.

Cooley, C. H. (1909). *Social organization: A study of the larger mind*. New York: Scribner's Sons.

Cosmides, L., Tooby, J. (2000). Evolutionary psychology and the emotions. In Lewis, R., Haviland-Jones, J. M. (Eds.), *Handbook of emotions*, 2nd ed. New York: Guilford, 91–115.

Clark, A. (2008). *Supersizing the mind: Embodiment, action, and cognitive extension*. New York: Oxford University Press.

Clay-Warner, J., Robinson, D. T. (2008) (Hg). *Social structure and emotion*. San Diego, CA: Academic Press.

Clore, G. L., Huntsinger, J. R. (2007). How emotions inform judgment and regulate thought. *Trends in Cognitive Sciences, 11(9)*, 393–399.

Clore, G. L., Ketelaar, T. (1997). Minding our emotions: On the role of automatic, unconscious affect. In Wyer, R. S. (Ed.), *The automaticity of everyday life*. Mahwah, NJ: Erlbaum, 105–120.

Clore, G. L., Ortony, A. (2000). Cognition in emotion: Always, sometimes, or never? In Lane, R. D., Nadel, L. (Eds.), *Cognitive neuroscience of emotion*. New York: Oxford University Press, 24–61.

Clore, G. L., Storbeck, J. (2006). Affect as information about liking, efficacy, and importance. In Forgas, J. P. (Ed.), *Affect in social thinking and behavior*. New York: Psychology Press, 123–142.

Clore, G. L., Schwarz, N., Conway, M. (1994). Affective causes and consequences of social information processing. In Wyer, R. S., Srull, T. K. (Eds.), *Handbook of social cognition*. Vol. 1, 2nd ed. Hillsdale, NJ: Erlbaum, 323–417.

Crenshaw, K. W. (1991). Mapping the margins: Intersectionality, identity politics, and violence against women of color. *Stanford Law Review, 43(6)*, 1241–1299.

Cynader, M. S., Frost, B. J. (1999). Mechanisms of brain development: Neuronal sculpting by the physical and social environment. In Keating, D.P., Hertzman, C. (Eds.), *Developmental health and the wealth of nations*. New York: Guilford, 153–184.

Dahrendorf, R. (1958). *Homo Sociologicus. Ein Versuch zur Geschichte, Bedeutung und Kritik der sozialen Rolle*. Opladen: Westdeutscher Verlag 1974.

Damasio, A. R. (1994). *Descartes' error*. New York: Quill/Harper Collins 2000.

Damasio, A. R. (2003). *Looking for Spinoza*. Orlando, FL: Harcourt.

Damasio, A. R., Grabowski, T. J., Bechara, A., Damasio, H., Ponto, L. L., Parvizi, J., Hichwa, R. D. (2000). Subcortical and cortical brain activity during the feeling of self-generated emotions. *Nature Neuroscience, 3(10)*, 1049–1056.

D'Andrade, R. G. (1981). The cultural part of cognition. *Cognitive Science, 5(3)*, 179–195.

Darwin, C. (1872). *The expression of the emotions in man and animals*. London: John Murray 1904.

Davidson, R. J. (2003a). Seven sins in the study of emotion: Correctives from affective neuro-science. *Brain and Cognition, 52(1)*, 129–132.

Davidson, R. J. (2003b). Darwin and the neural bases of emotion and affective style. *Annals of the New York Academy of Sciences, 1000*, 316–336.

Davidson, R. J. (2003c). Affective neuroscience and psychophysiology: Toward a synthesis. *Psychophysiology, 40(5)*, 655–665.

Davidson, R. J. (2004). What does the prefrontal cortex "do" in affect: perspectives on frontal EEG asymmetry research. *Biological Psychology, 67(1/2)*, 219–233.

Davidson, R. J., Irwin, W. (1999). The functional neuroanatomy of emotion and affective style. *Trends in Cognitive Sciences, 3(1)*, 11–21.

Davidson, R. J., Jackson, D. C., Kalin, N. H. (2000). Emotion, plasticity, context, and regulation: Perspectives from affective neuroscience. *Psychological Bulletin, 126(6)*, 890–909.

De Houwer, J., Hermans, D. (2001). Editorial: Automatic affective processing. *Cognition and Emotion, 15(2)*, 113–114.

Denzin, N. K. (1980). A phenomenology of emotion and deviance. *Zeitschrift für Soziologie, 9(3)*, 251–261.

Denzin, N. K. (1984). *On understanding emotion*. San Francisco, CA: Jossey-Bass.

DiMaggio, P. (1991). The micro–macro dilemma in organizational research: Implicat-ions of role-system theory. In Huber, J. (Ed.), *Macro–micro linkages in sociology*. Newbury Park, CA: Sage, 76–98.

DiMaggio, P. (1997). Culture and cognition. *Annual Review of Sociology*, *23*, 263–287.

DiMaggio, P. (2002). Why cognitive (and cultural) sociology needs cognitive psychology. In Cerulo, K. A. (Ed.), *Culture in mind*. New York: Routledge, 274–282.

Dimberg, U., Thunberg, M. (1998). Rapid facial reactions to emotional facial expressions. *Scandinavian Journal of Psychology*, *39(1)*, 39–45.

Döveling, K., Schwarz, C. (2010). Politics in the living room. The "democratization of fame" in Pop Idol formats. In Baruh, L., Park, J. H. (Eds.), *Reel politics: Reality television as a platform for political discourse*. Cambridge: Cambridge Scholars Publishing, 95–115.

Dunbar, R. I. (2002). The social brain hypothesis. In Cacioppo, J. T., Berntson, G. G., Adolphs, R., Carter, C. S., Davidson, R. J., McClintock, M. K., McEwen, B. S., Meaney, M. J., Schacter, D. L., Sternberg, E. M., Suomi, S. S., Taylor, S. E. (Eds.), *Foundations in social neuroscience*. Cambridge, MA: MIT Press, 69–88.

Duncan, S., Barrett, L. (2007). Affect is a form of cognition: A neurobiological analysis. *Cognition and Emotion*, *21(6)*, 1184–1211.

Dunn, B. D., Dalgleish, T., Lawrence A. D. (2006). The somatic marker hypothesis: A critical evaluation. *Neuroscience and Biobehavioral Reviews*, *30*, 239–271.

Durkheim, E. (1915). *The Elementary Forms of the Religious Life*. London: Allen & Unwin.

Durkheim, E. (1938). *The rules of sociological method*. Chicago, IL: University of Chicago Press.

Eder, A. B., Hommel, B., De Houwer, J. (2007). How distinctive is affective processing? On the implications of using cognitive paradigms to study affect and emotion. *Cognition and Emotion*, *21(6)*, 1137–1154.

Elfenbein, H. A., Beaupré, M., Lévesque, M., Hess, U. (2007). Toward a dialect theory: Cultural differences in the expression and recognition of posed facial expressions. *Emotion*, *7*, 131–146.

Elfenbein, H. A., Polzer, J. T., Ambady, N. (2007). Team emotion recognition accuracy and team performance. In Ashkanasy, N. M., Zerbe, W. J., Härtel, C. E. (Eds.), *Research on emotions in organizations*. Vol. 3. Oxford: Elsevier, 87–119.

Ehrenberg, A. (2010). *The weariness of the self: Diagnosing the history of depression in the contemporary age*. Montreal: McGill-Queen's University Press.

Ehrenreich, B. (2009). *Bright-sided: How the relentless promotion of positive thinking has undermined America*. New York: Holt.

Eid, M., Diener, E. (2001). Norms for experiencing emotions in different cultures: Inter- and intra-national differences. *Journal of Personality and Social Psychology*, *81*, 869–885.

Eisenberg, L. (1995). The social construction of the human brain. *American Journal of Psychiatry*, *152(11)*, 1563–1575.

Eisenberger, N. I., Lieberman, M. D. (2004). Why rejection hurts: A common neural alarm system for physical and social pain. *Trends in Cognitive Sciences*, *8(7)*, 294–300.

Eisenberger, N. I., Lieberman, M. D., Williams, K. D. (2003). Does rejection hurt? An fMRI study of social exclusion. *Science*, *302*, 290–292.

Ekman, P. (1972). Universals and cultural differences in facial expressions of emotion. In Cole, J. (Ed.), *Nebraska symposium on motivation*. Vol. 19. Lincoln, NE: University of Nebraska Press, 207–282.

Ekman, P. (1982). *Emotion in the human face.* Cambridge: Cambridge University Press.

Ekman, P. (1992a). An argument for basic emotions. *Cognition and Emotion, 6(3/4),* 169–200.

Ekman, P. (1992b). Facial expression of emotion. New findings, new questions. *Psychological Science, 3(1),* 34–38.

Ekman, P., Friesen, W. V. (1975). *Unmasking the face: A guide to recognizing emotions from clues.* Upper Saddle River, NJ: Prentice Hall.

Elbert, T., Heim, S., Rockstroh, B. (2001). Neural plasticity and development. In Nelson, C. A., Luciana, M. (Eds.), *Handbook of developmental cognitive neuroscience.* Cambridge, MA: MIT Press, 191–202.

Elfenbein, H. A., Ambady, N. (2002). Is there an in-group advantage in emotion recognition? *Psychological Bulletin, 128(2),* 243–249.

Elfenbein, H. A., Ambady, N. (2003a). Universals and cultural differences in recognizing emotions. *Current Directions in Psychological Science, 12(5),* 159–164.

Elfenbein, H. A., Ambady, N. (2003b). When familiarity breeds accuracy: Cultural exposure and facial emotion recognition. *Journal of Personality and Social Psychology, 85(2),* 276–290.

Elfenbein, H. A., Mandal, M. K., Ambady, N., Harizuka, S. (2002). Cross-cultural patterns in emotion recognition: Highlighting design and analytical techniques. *Emotion, 2(1),* 75–84.

Elias, N. (1994). *The civilizing process. Sociogenetic and psychogenetic investigations.* Oxford: Blackwell.

Ellickson, R. C. (1991). *Order without law. How neighbors settle disputes.* Cambridge, MA: Harvard University Press.

Ellsworth, P. C. (1994). Levels of thought and levels of emotion. In Ekman, P., Davidson, R. J. (Eds.), *The nature of emotion.* New York: Oxford University Press, 192–196.

Elster, J. (1989). Social norms and economic theory. *Journal of Economic Perspectives, 3(4),* 99–117.

Elster, J. (1996). Rationality and the emotions. *The Economic Journal, 106(438),* 1386–1397.

Elster, J. (1999). *Alchemies of the mind. Rationality and the emotions.* New York: Cambridge University Press.

Elster, J. (2004). Emotions and rationality. In Manstead, A. S., Frijda, N. H., Fischer, A. (Eds.), *Feelings and emotions.* New York: Oxford University Press, 30–48.

Engelen, E.-M., Markowitsch, H. J., von Scheve, C., Röttger-Rössler, B., Stephan, A., Holodynski, M., Vandekerckhove, M. (2008). Emotions as bio-cultural processes: Disciplinary debates and an interdisciplinary outlook. In Röttger-Rössler, B., Markowitsch, H. J. (Eds.), *Emotions as bio-cultural processes.* New York: Springer

Erk, S., Spitzer, M., Wunderlich, A., Galley, L., Walter, H. (2002). Cultural objects modulate reward circuitry. *NeuroReport, 13(18),* 2499–2503.

Evans, D. (2002). The search hypothesis of emotions. *British Journal for the Philosophy of Science, 53(4),* 497–509.

Evans, J. S. (2008). Dual-processing accounts of reasoning, judgment, and social cognition. *Annual Review of Psychology, 59,* 255–278.

Falk, A., Fehr, E., Fischbacher, U. (2005). Driving forces behind informal sanctions. *Econometrica, 73(6),* 2017–2030.

Fehr, E., Fischbacher, U. (2004a). Social norms and human cooperation. *Trends in Cognitive Sciences, 8(4),* 185–190.

Fehr, E., Fischbacher, U. (2004b). Third-party punishment and social norms. *Evolution and Human Behavior*, *25(1)*, 63–87.

Fehr, E., Gächter, S. (2002). Altruistic punishment in humans. *Nature*, *415*, 137–140.

Fiedler, K., Bless, H. (2000). The formation of beliefs at the interface of affective and cognitive processes. In Frijda, N. H., Manstead, A., Bem, S. (Eds.), *Emotions and beliefs. How feelings influence thoughts*. Cambridge: Cambridge University Press, 144–170.

Fiehler, R. (1990). *Kommunikation und Emotion*. Berlin de Gruyter.

Fineman, S. (2003). *Understanding emotion at work*. London: Sage.

Fischer, A., Rotteveel, M., Manstead, A. S. (2004). Emotional assimilation: How we are influenced by others' emotions. *Cahiers de Psychologie Cognitive*, *22(2)*, 223–246.

Fiske, S. T. (1982). Schema-triggered affect: Applications to social perception. In Clark, M. S., Fiske, S. T. (Eds.), *Affect and cognition*. Hillsdale, NJ: Erlbaum, 55–78.

Fiske, S. T., Taylor, S. E. (1984). *Social cognition*. New York: Random House.

Flam, H. (1990). Emotional man: I. The emotional man and the problem of collective action. *International Sociology*, *5(1)*, 39–56.

Flam, H. (1998). *Mosaic of fear. Poland and East Germany before 1989*. New York: Columbia University Press.

Fontaine, J. R., Scherer, K. R., Roesch, E. B., Ellsworth, P. C. (2007). The world of emotions is not two-dimensional. *Psychological Science*, *18*, 1050–1057.

Forgas, J. P. (1995). Mood and judgment: The affect infusion model (AIM). *Psychological Bulletin*, *117(1)*, 39–66.

Forgas, J. P. (2000). Affect and information processing strategies: An interactive relationship. In Forgas, J. P. (Ed.), *Feeling and thinking. The role of affect in social cognition*. New York: Cambridge University Press, 253–282.

Forgas, J. P. (2006). Affective influences on interpersonal behavior: Towards understanding the role of affect in everyday interactions. In Forgas, J. P. (Ed.), *Affect in social thinking and behavior*. New York: Psychology Press, 269–290.

Frank, R. H. (1988). *Passions within reason*. New York: Norton.

Frank, R. H. (1993). The strategic role of emotions: Reconciling over and undersocialized accounts of behaviour. *Rationality and Society*, *5(2)*, 160–184.

Franks, D. D. (2010). *Neurosociology*. New York: Springer.

Franks, D. D., Smith, T. S. (Eds.) (1999). *Mind, brain, and society: Toward a neurosociology of emotion*. Greenwich, CT: JAI Press.

Freese, J., Li, J.-C., Wade, L. D. (2003). The potential relevances of biology to social inquiry. *Annual Review of Sociology*, *29*, 233–256.

Frevert, U. (2011). *Emotions in history: Lost and found*. Budapest: Central European University Press.

Fridlund, A. J. (1994). *Human facial expression: An evolutionary view*. San Diego, CA: Academic Press.

Frijda, N. H. (1986). *The emotions*. Cambridge: Cambridge University Press.

Frijda, N. H. (1994). Emotions require cognitions, even if simple ones. In Ekman, P., Davidson, R. J. (Eds.), *The nature of emotion*. New York: Oxford University Press, 197–202.

Frijda, N. H. (2004). Emotions and action. In Manstead, A. S., Frijda, N. H., Fischer, A. (Eds.), *Feelings and emotions*. New York: Oxford University Press, 158–173.

Frijda, N. H., Mesquita, B. (1994). The social roles and functions of emotions. In Kitayama, S., Markus, H. R. (Eds.), *Emotion and culture*. Washington, DC: American Psychological Association, 51–88.

Frijda, N., Zeelenberg, M. (2001). Appraisal. What is the dependent? In Scherer, K. R., Schorr, A., Johnstone, T. (Eds.), *Appraisal processes in emotion.* New York: Oxford University Press, 141–155.

Frijda, N. H., Manstead, A., Bem, S. (2000). The influence of emotions on beliefs. In Frijda, N. H., Manstead, A., Bem, S. (Eds.), *Emotions and beliefs. How feelings influence thoughts.* Cambridge: Cambridge University Press, 1–9.

Gainotti, G. (2000). Neuropsychological theories of emotion. In Borod, J. C. (Ed.), *The neuropsychology of emotion.* New York: Oxford University Press, 214–236.

Gallo, L. C., Matthews, K. A. (2003). Understanding the association between socioeconomic status and physical health: Do negative emotions play a role? *Psychological Bulletin, 129(1),* 10–51.

Giddens, A. (1984). *The constitution of society.* Berkeley, CA: University of California Press.

Gigerenzer, G. (2007). *Gut feelings: The intelligence of the unconscious.* New York: Viking.

Gigerenzer, G., Gaissmaier, W. (2011). Heuristic decision-making. *Annual Review of Psychology, 62,* 451–482.

Glimcher, P. W. (2003). *Decisions, uncertainty, and the brain.* Cambridge, MA: MIT Press.

Goffman, E. (1959). *The presentation of self in everyday life.* New York: Doubleday.

Goldie, P. (2000). *The emotions.* New York: Oxford University Press.

Goldie, P. (2004). Emotion, feeling, and knowledge of the world. In Solomon, R. C. (Ed.), *Thinking about feeling.* New York: Oxford University Press, 91–106.

Goleman, D. (1995). *Emotional intelligence.* New York: Bantam.

Grandey, A. A. (2000). Emotion regulation in the workplace: A new way to conceptualize emotional labor. *Journal of Occupational Health Psychology, 5(1),* 95–110.

Grandey, A. A., Brauburger, A. (2002). The emotion regulation behind the customer service smile. In Lord, R., Klimoski, R., Kanfer, R. (Eds.), *Emotions in the workplace.* San Francisco, CA: Jossey-Bass, 260–294.

Gray, J. A. (1990). Brain systems that mediate both emotion and cognition. *Cognition and Emotion, 4(3),* 269–288.

Greco, M., Stenner, P. (Eds.) (2007). *Emotions: A social science reader.* London: Routledge.

Griffiths, P. E. (1997). *What emotions really are.* Chicago, IL: University of Chicago Press.

Gross, J. J. (1998). The emerging field of emotion regulation: An integrative review. *Review of General Psychology, 2(3),* 271–299.

Gross, J. J. (1999a). Emotion regulation: Past, present, future. *Cognition and Emotion, 13(5),* 551–573.

Gross, J. J. (1999b). Emotion and emotion regulation. In Pervin, L. A., John, O. P. (Eds.), *Handbook of personality. Theory and research.* 2nd ed. New York: Guilford, 525–552.

Gross, J. J. (2002). Emotion regulation: Affective, cognitive, and social consequences. *Psychophysiology, 39(3),* 281–291.

Gross, J. J., Barrett, L. F. (2011). Emotion generation and emotion regulation: One or two depends on your point of view. *Emotion Review, 3(1),* 8–16.

Gross, J. J., John, O. P. (2002). Wise emotion regulation. In Barrett, L., Salovey, P. (Eds.), *The wisdom in feeling.* New York: Guilford, 297–318.

Güth, W., Schmittberger, R., Schwarze, B. (1982). An experimental analysis of ultimatum bargaining. *Journal of Economic Behavior and Organization, 3*, 367–88.

Hackman, D., Farah, M. J. (2009). Socioeconomic status and brain development. *Trends in Cognitive Sciences, 13*, 65–73.

Hackman, D. A., Farah, M. J., Meaney, M. J. (2010). Socioeconomic status and the brain: Mechanistic insights from human and animal research. *Nature Reviews Neuroscience, 11*, 651–659.

Haidt, J., Joseph, C. (2004). Intuitive ethics: How innately prepared intuitions generate culturally variable virtues. *Daedalus, 133*, 55–66.

Haidt, J., Keltner, D. (1999). Culture and facial expression: Open-ended methods find more expressions and a gradient of recognition. *Cognition and Emotion, 13(3)*, 225–266.

Hamann, S. (2001). Cognitive and neural mechanisms of emotional memory. *Trends in Cognitive Sciences, 5(9)*, 394–400.

Hammond, M. (1990). Affective maximization. A new macro-theory in the sociology of emotion. In Kemper, T. D. (Ed.), *Research agendas in the sociology of emotions*. Albany, NY: State University of New York Press, 58–81.

Hammond, M. (2003). The enhancement imperative: The evolutionary neurophysiology of Durkheimian solidarity. *Sociological Theory, 21(4)*, 359–374.

Harding, J., Pribram, E. D. (Eds.) (2009). *Emotions: A cultural studies reader*. London: Routledge.

Haxby, J. V., Hoffman, E. A., Gobbini, M. I. (2000). The distributed human neural system for face perception. *Trends in Cognitive Sciences, 4(6)*, 223–233.

Hatfield, E., Cacioppo, J. T., Rapson, R. L. (1992). Primitive emotional contagion. In Clark, M. S. (Ed.), *Emotion and social behavior*. Newbury Park, CA: Sage, 151–177.

Hatfield, E., Cacioppo, J. T., Rapson, R. L. (1994). *Emotional contagion*. New York: Cambridge University Press.

Hechter, M., Opp, K.-D. (Eds.) (2001). *Social norms*. New York: Russell Sage.

Heise, D. R. (1979). *Understanding events. Affect and the construction of social action*. New York: Cambridge University Press.

Heise, D. R. (2010). *Surveying cultures: Discovering shared conceptions and sentiments*. New York: Wiley.

Heise, D. R., Calhan, C. (1995). Emotion norms in interpersonal events. *Social Psychology Quarterly, 58(4)*, 223–240.

Henrich, J., Boyd, R. (1998). The evolution of conformist transmission and the emergence of between-group differences. *Evolution and Human Behavior, 19*, 215–241.

Hess, U., Thibault, P. (2009). Darwin and emotion expression. *American Psychologist, 64*, 120–128.

Hess, U., Banse, R., Kappas, A. (1995). The intensity of facial expression is determined by underlying affective state and social situation. *Journal of Personality and Social Psychology, 69(2)*, 280–288.

Hess, U., Philippot, P., Blairy, S. (1998). Facial reactions to emotional facial expressions: Affect or cognition? *Cognition and Emotion, 12(4)*, 509–531.

Hinson, J. M., Jameson, T. L., Whitney, P. (2002). Somatic markers, working memory, and decision-making. *Cognitive, Affective, Behavioral Neuroscience, 2(4)*, 341–353.

Hochschild, A. R. (1979). Emotion work, feeling rules, and social structure. *American Journal of Sociology, 85(3)*, 551–575.

Hochschild, A. R. (1983). *The managed heart*. Berkeley, CA: University of California Press.

Holodynski, M., Friedlmeier, W. (2005). *Development of emotions and emotion regulation.* New York: Springer.

Homans, G. C. (1950). *The human group.* New York: Harcourt, Brace.

Hornak, J., Bramham, J., Rolls, E. T., Morris, R. G., O'Doherty, J., Bullock, P. R., Polkey, C. E. (2003). Changes in emotion after circumscribed surgical lesions of the orbitofrontal and cingulate cortices. *Brain, 126,* 1691–1712.

Horne, C. (2001). Sociological perspectives on the emergence of norms. In Hechter, M., Opp, K.-D. (Eds.), *Social norms.* New York: Russell Sage, 3–34.

Horne, C., Cutlip, A. (2002). Sanctioning costs and norm enforcement: An experimental test. *Journal of Rationality and Society, 14(3),* 285–307.

House, J. S. (1981). Social structure and personality. In Rosenberg, M., Turner, R. H. (Eds.), *Social psychology. Sociological perspectives.* New York: Basic Books, 525–561.

Howard, J. A. (1991). Introduction: The self–society dynamic. In Howard, J. A., Callero, P. L. (Eds.), *The self–society dynamic.* New York: Cambridge University Press, 1–17.

Howard, J. A. (1995). Social cognition. In Cook, K. S., Fine, G. A., House, J. S. (Eds.), *Sociological perspectives on social psychology.* Boston, MA: Allyn & Bacon, 90–117.

Hutchins, E. (1996). *Cognition in the wild.* Cambridge, MA: MIT Press.

Hutchins, E. (1991). The social organization of distributed cognition. In Resnick, L. B., Levine, J. M., Teasley, S. D. (Eds.), *Perspectives on socially shared cognition.* Washington, DC: American Psychological Association, 283–307.

Illouz, E. (1997). *Consuming the romantic utopia. Love and the cultural contradictions of capitalism.* Berkeley, CA: University of California Press.

Illouz, E. (2007). *Cold intimacies: The making of emotional capitalism.* Malden, MA: Blackwell.

Illouz, E. (2008). *Saving the modern soul.* Berkeley, CA: University of California Press.

Izard, C. E. (1971). *The face of emotion.* New York: Appleton-Century-Crofts.

Izard, C. E. (1977). *Human emotions.* New York: Plenum.

James, W. (1884). What is an emotion? *Mind, 9(34),* 188–205.

James, W. (1897). The sentiment of rationality. In James, W. (Ed.), *The will to believe and other essays in popular philosophy.* New York: Longman, Green & Co, 63–110.

Goodwin, J., Jasper, J. M., Polletta, F. (2004). Emotional dimensions of social movements. In Snow, D. A., Soule, S. A., Kriesi, H. (Eds.), *The Blackwell companion to social movements.* Malden, MA: Blackwell, 413–432.

Johnson-Laird, P. N., Oatley, K. (1992). Basic emotions, rationality, and folk theory. *Cognition and Emotion, 6(3/4),* 201–223.

Kappas, A. (2002a). What facial activity can and cannot tell us about emotions. In Katsikitis, M. (Ed.), *The human face. Measurement and meaning.* Dordrecht: Kluwer, 215–234.

Kappas, A. (2002b). The science of emotion as a multidisciplinary research paradigm. *Behavioural Processes, 60(2),* 85–98.

Kappas, A. (2008). Psssst! Dr. Jekyll and Mr. Hyde are actually the same person! A tale of regulation and emotion. In Vandekerckhove, M., von Scheve, C., Ismer, S., Jung, S., Kronast, S. (Eds.), *Regulating emotions. Culture, social necessity and biological inheritance.* Malden, MA: Wiley-Blackwell, 15–38.

Kappas, A. (2011). Emotion and regulation are one! *Emotion Review, 3(1),* 17–25.

Keltner, D., Haidt, J. (1999). Social functions of emotion at four levels of analysis. *Cognition and Emotion, 13(5),* 505–521.

Keltner, D., Ekman, P., Gonzaga, G. C., Beer, J. (2003). Facial expression of emotion. In Davidson, R. J., Scherer, K. R., Goldsmith, H. H. (Eds.), *Handbook of affective sciences*. New York: Oxford University Press, 415–432.

Kemper, T. D. (1978). *A social interactional theory of emotions*. New York: Wiley.

Kemper, T. D. (1981). Social constructionist and positivist approaches to the sociology of emotions. *American Journal of Sociology, 87(2)*, 336–362.

Kemper, T. D. (1984). Power, status, and emotions. A sociological contribution to a psychophysiological domain. In Scherer, K. R., Ekman, P. (Eds.), *Approaches to emotion*. Hillsdale, NJ: Erlbaum, 369–383.

Ketelaar, T., Au, W. T. (2003). The effects of guilty feelings on the behavior of uncooperative individuals in repeated social bargaining games: An Affect-as-information interpretation of the role of emotion in social interaction. *Cognition and Emotion, 17(3)*, 429–453.

Knorr-Cetina, K. D. (1981). Introduction: The micro-sociological challenge of macro-sociology. In Knorr-Cetina, K. D., Cicourel, A. V. (Eds.), *Advances in social theory and methodology. Toward an integration of micro and macro-sociologies*. Boston, MA: Routledge & Kegan Paul, 1–47.

Knutson, B., Bossaerts, P. (2007). Neural antecedents of financial decisions. *The Journal of Neuroscience, 27(31)*, 8174–8177.

Kober, H., Barrett, L. F., Joseph, J., Bliss-Moreau, E., Lindquist, K. A., Wager, T. D. (2008). Functional networks and cortical-subcortical interactions in emotion: A meta-analysis of neuroimaging studies. *NeuroImage, 42*, 998–1031.

Kolb, B., Whishaw, I. Q. (1998). Brain plasticity and behavior. *Annual Review of Psychology, 49*, 43–64.

Kringelbach, M. L., Rolls, E. T. (2004). The functional neuroanatomy of the human OFC: evidence from neuroimaging and neuropsychology. *Progress in Neurobiology, 72(5)*, 341–372.

Kuhnen, C. M., Knutson, B. (2005). The neural basis of financial risk taking. *Neuron, 47*, 763–770.

LaBar, K. S., Cabeza, R. (2006). Cognitive neuroscience of emotional memory. *Nature Reviews Neuroscience, 7*, 54–64.

Lahlou, S. (2001). Functional aspects of social representation. In Deaux, K., Philogène, G. (Eds.), *Representations of the social*. Oxford: Blackwell, 131–146.

Lawler, E. J. (2001). An affect theory of social exchange. *American Journal of Sociology, 107(2)*, 321–52.

Lawler, E. J., Thye, S. R. (1999). Bringing emotion into social exchange theory. *Annual Review of Sociology, 25*, 217–244.

Lazarus, R. S. (1966). *Psychological stress and the coping process*. New York: McGraw-Hill.

Lazarus, R. S. (1968). Emotion and adaptation: Conceptual and empirical relations. In Arnold, W. J. (Ed.), *Nebraska symposium on motivation*. Lincoln, NE: University of Nebraska Press, 175–266.

Lazarus, R. S. (1984). Thoughts on the relations between emotion and cognition. In Scherer, K. R., Ekman, P. (Eds.), *Approaches to emotion*. Hillsdale, NJ: Erlbaum, 247–257.

Lazarus, R. S. (1991a). *Emotion and adaptation*. New York: Oxford University Press.

Lazarus, R. S. (1991b). Progress on a cognitive-motivational-relational theory of emotion. *American Psychologist, 46(8)*, 819–834.

Lazarus, R. S., Smith, C. A. (1988). Knowledge and appraisal in the cognition-emotion relationship. *Cognition and Emotion, 2(4)*, 281–300.

Le Bon, G. (1896). *The crowd*. New York: Macmillan.

LeDoux, J. E. (1993). Cognition versus emotion, again—this time in the brain: A response to Parrott and Schulkin. *Cognition and Emotion, 7(1)*, 61–64.

LeDoux, J. E. (1995). Emotion: Clues from the brain. *Annual Review of Psychology, 46*, 209–235.

LeDoux, J. E. (1996). *The emotional brain*. New York: Touchstone.

LeDoux, J. E. (2000). Emotion circuits in the brain. *Annual Review of Neuroscience, 23*, 155–184.

Lerner, J. S., Keltner, D. (2000). Beyond valence: Toward a model of emotion-specific influences on judgement and choice. *Cognition and Emotion, 14(4)*, 473–493.

Levenson, R. W. (1999). The intrapersonal functions of emotion. *Cognition and Emotion, 13(5)*, 481–504.

Levenson, R. W. (2003). Autonomic specificity and emotion. In Davidson, R. J., Scherer, K. R., Goldsmith, H. H. (Eds.), *Handbook of affective sciences*. New York: Oxford University Press, 212–224.

Leventhal, H., Scherer, K. R. (1987). The relationship of emotion to cognition. A functional approach to a semantic controversy. *Cognition and Emotion, 1(1)*, 3–28.

Lewis, M. D. (2005). Bridging emotion theory and neurobiology through dynamic systems theory. *Behavioral and Brain Sciences, 28(2)*, 169–194.

Leyens, J.-P., Dardenne, B. (1996). Soziale Kognition. Ansätze und Grundbegriffe. In Stroebe, W., Hewstone, M., Stephenson, G. M. (Eds.), *Sozialpsychologie. Eine Einführung*. 3rd ed. Berlin: Springer, 116–141.

Lieberman, M. D. (2007). Social cognitive neuroscience: A review of core processes. *Annual Review of Psychology, 58*, 259–289.

Lieberman, M. D., Eisenberger, N. I., Crockett, M. J., Tom, S. M., Pfeifer, J. H. und B. M. Way (2007). Putting feelings into words: Affect labeling disrupts amygdala activity in response to affective stimuli. *Psychological Science, 18(5)*, 421–428.

Lindquist, K. A., Wager, T. D., Kober, H., Bliss-Moreau, E., L. F. Barrett (2012). The brain basis of emotion: A meta-analytic review. *Behavioral and Brain Sciences, 35(3)*, 121–143.

Lizardo, O. (2004). The cognitive origins of Bourdieu's *habitus*. *Journal for the Theory of Social Behaviour, 34(4)*, 375–401.

Loewenstein, G., Lerner, J. S. (2003). The role of affect in decision-making. In Davidson, R. J., Scherer, K. R., Goldsmith, H. H. (Eds.), *Handbook of affective sciences*. New York: Oxford University Press, 619–642.

Loewenstein, G., Weber, E. U., Hsee, C. K., N. Welch (2001). Risk as feelings. *Psychological Bulletin, 127(2)*, 267–286.

Lundqvist, L. O., Dimberg, U. (1995). Facial expressions are contagious. *Journal of Psychophysiology, 9*, 203–211.

McGaugh, J. L. (2003). *Memory and emotion*. New York: Columbia University Press.

MacLean, P. (1952). Some psychiatric implications of physiological studies on frontotemporal portion of limbic system (visceral brain). *Electroencephalography and Clinical Neurophysiology, 4(4)*, 407–418.

Macmillan, M. (2000). *An odd kind of fame: Stories of Phineas Gage*. Cambridge, MA: MIT Press.

Macrae, C. N., Bodenhausen, G. V. (2000). Social cognition: Thinking categorically about others. *Annual Review of Psychology*, *51*, 93–120.

Macrae, C. N., Bodenhausen, G. V. (2001). Social cognition: Categorical person perception. *British Journal of Psychology*, *92(1)*, 239–255.

Mannheim, K. (1936). *Ideology and utopia*. London: Routledge.

Manstead, A. S., Fischer, A. H. (2001). Social appraisal: The social world as object of and influence on appraisal processes. In Scherer, K. R., Schorr, A., Johnstone, T. (Eds.), *Appraisal processes in emotion*. New York: Oxford University Press, 221–232.

Marby, J. B. (1999). *Social structure and anger: Social psychological mediators*. Ph.D. Dissertation, Virginia Polytechnic Institute and State University. Blacksburg, VA.

Markowitsch, H. J. (1999). *Cognitive neuroscience of memory*. Göttingen: Hogrefe & Huber.

Marmot, M. (2004). *The status syndrome*. New York: Times Books.

Marsh, A. A., Elfenbein, H. A., Ambady, N. (2003). Nonverbal "accents": Cultural differences in facial expressions of emotion. *Psychological Science*, *14(4)*, 373–376.

Martin, J. L. (2011). *Social structures*. Princeton, NJ: Princeton University Press.

Massey, D. (2008). *Categorically unequal: The American stratification system*. New York: Russell Sage Foundation.

Matsumoto, D., Seung H. Y., J. Fontaine (2008). Mapping expressive differences around the world: The relationship between emotional display rules and individualism versus collectivism. *Journal of Cross-cultural Psychology*, *39*, 55–74.

Mauss, I. B., Bunge, S. A., Gross, J. J. (2008). Culture and automatic emotion regulation. In Vandekerckhove, M., von Scheve, C., Ismer, S., Jung, S., Kronast, S. (Eds.), *Regulating emotions. Culture, social necessity and biological inheritance*. Malden, MA: Wiley-Blackwell, 39–60.

Mead, G. H. (1934). *Mind, self, and society*. Chicago, IL: University of Chicago Press.

Morris, J. S., Öhman, A., Dolan, R. J. (1998). Conscious and unconscious emotional learning in the human amygdala. *Nature*, *393*, 467–470.

Moscovici, S. (1961). *La Psychoanalyse son Image et son Public*. 2nd ed. Paris: Presses Universitaires de France.

Moscovici, S. (2001). Why a theory of social representations? In Deaux, K., Philogène, G. (Eds.), *Representations of the social*. Oxford: Blackwell, 8–36.

Mowrer, S. M., Jahn, A. A., Abduljalil, A., Cunningham, W. A. (2011). The value of success: acquiring gains, avoiding losses, and simply being successful. *PLoS One*, *6(9)*, e25307.

Mummenthaler, C., Sander, D. (2012). Social appraisal influences recognition of emotions. *Journal of Personality and Social Psychology*, *102(6)*, 1118–1135.

Murray, E. A. (2007). The amygdala, reward and emotion. *Trends in Cognitive Sciences*, *11(11)*, 489–497.

Neckel, S. (1991). *Status und Scham*. Frankfurt am Main/New York: Campus.

Neckel, S. (1999). Blanker Neid, blinde Wut? Sozialstruktur und kollektive Gefühle. *Leviathan*, *27(2)*, 145–165.

Neckel, S. (2005). Emotion by design: Self-management of feelings as a cultural program. In Röttger-Rössler, B., Markowitsch, H. J. (Eds.), *Emotions as bio-cultural processes*. New York: Springer, 181–198.

Newton, T. (2003). Truly embodied sociology: marrying the social and the biological? *The Sociological Review*, *51(1)*, 20–42.

Nichols, S. (2004). *Sentimental rules: On the natural foundations of moral judgment.* New York: Oxford University Press.

Noë, A. (2004). *Action in perception.* Cambridge, MA: MIT Press.

Noble, K. G., McCandliss, B. D., Farah, M. J. (2007). Socioeconomic gradients predict individual differences in neurocognitive abilities. *Developmental Science, 10(4),* 464–480.

Oatley, K. (1992). *Best laid schemes.* New York: Cambridge University Press.

Oatley, K. (2000). The sentiments and beliefs of distributed cognition. In Frijda, N. H., Manstead, A. S., Bem, S. (Eds.), *Emotions and beliefs. How feelings influence thoughts.* Cambridge: Cambridge University Press, 78–107.

Ochsner, K. N., Barrett, L. F. (2001). A multiprocess perspective on the neuroscience of emotion. In Mayne, T. J., Bonnano, G. (Eds.), *Emotion: Current issues and future directions.* New York: Guilford, 38–81.

Ochsner, K. N., Gross, J. J. (2004). Thinking makes it so. A social cognitive neuroscience approach to emotion regulation. In Baumeister, R., Vohs, K. (Eds.), *The handbook of self-regulation.* New York: Guilford, 229–255.

Ochsner, K. N., Gross, J. J. (2005). The cognitive control of emotion. *Trends in Cognitive Sciences, 9(5),* 242–249.

Ochsner, K. N., Bunge, S. A., Gross, J. J., Gabrieli, J. D. (2002). Rethinking feelings: An fMRI study of the cognitive regulation of emotion. *Journal of Cognitive Neuroscience, 14(8),* 1215–1229.

O'Doherty, J., Kringelbach, M. L., Rolls, E. T., Hornak, J., Andrews, C. (2001). Abstract reward and punishment representations in the human OFC. *Nature Neuroscience, 4(1),* 95–102.

Öhman, A., Flykt, A., Lundqvist, D. (2000). Unconscious emotion: Evolutionary perspectives, psychophysiological data and neuropsychological mechanisms. In Lane, R. D., Nadel, L. (Eds.), *Cognitive neuroscience of emotion.* New York: Oxford University Press, 296–327.

Oliver, M. B., Woolley, J. K. (2010). Tragic and poignant entertainment: The gratifications of meaningfulness. In Döveling, K., von Scheve, C., Konijn, E. (Eds.), *The Routledge handbook of emotions and mass media.* New York: Routledge, 134–147.

Olsson, A., Ebert, J.P., Banaji, M. R., Phelps, E. A. (2005). The role of social groups in the persistence of learned fear. *Science, 309,* 785–787.

Opp, K.-D. (2002). When do norms emerge by human design and when by the unintended consequences of human action? The example of the no-smoking norm. *Rationality and Society, 14(2),* 131–158.

Ortony, A., Turner, T. J. (1990). What's basic about basic emotions? *Psychological Review, 97(3),* 315–331.

Ortony, A., Clore, G. L., Collins, A. (1988). *The cognitive structure of emotions.* New York: Cambridge University Press.

Osgood, C. E., Suci, G. J., Tannenbaum, P. H. (1957). *The measurement of meaning.* Urbana, IL: University of Illinois Press.

Panksepp, J. (1998). *Affective neuroscience.* New York: Oxford University Press.

Panksepp, J. (2003a). At the interface of the affective, behavioral, and cognitive neurosciences: Decoding the emotional feelings of the brain. *Brain and Cognition, 52(1),* 4–14.

Panksepp, J. (2003b). Damasio's error? Review of "Looking for Spinoza: Joy, sorrow, and the feeling brain" by A. Damasio. *Consciousness and Emotion, 4(1),* 111–134.

Papez, J. W. (1937). A proposed mechanism of emotion. *Archives of Neurology and Psychiatry, 79*, 217–224.

Pareto, V. (1935). *The mind and society.* New York: Harcourt Brace.

Parkinson, B. (1995). *Ideas and realities of emotion.* London: Routledge

Parkinson, B. (1997). Untangling the appraisal–emotion connection. *Personality and Social Psychology Review, 1(1)*, 62–79.

Parkinson, B. (2011). Interpersonal emotion transfer: Contagion and social appraisal. *Social and Personality Psychology Compass, 5(7)*, 428–439.

Parkinson, B., Manstead, A. S. (1992). Appraisal as a cause of emotion. In Clark, M. S. (Ed.), *Emotion.* Newbury Park, CA: Sage, 122–149.

Parkinson, B., Simons, G. (2009). Affecting others: social appraisal and emotion contagion in everyday decision making. *Personality and Social Psychology Bulletin, 35(8)*, 1071–1084.

Parrott, W. G., Schulkin, J. (1993). Neuropsychology and the cognitive nature of the emotions. *Cognition and Emotion, 7(1)*, 43–59.

Parsons, T. (1951). *The social system.* New York: Free Press.

Phan, K. L., Wager, T., Taylor, S. F., I. Liberzon (2002). Functional neuroanatomy of emotion: A meta-analysis of emotion activation studies in PET and fMRI. *NeuroImage, 16(2)*, 331–348.

Phelps, E. A. (2004). Human emotion and memory: interactions of the amygdala and hippocampal complex. *Current Opinion in Neurobiology, 14(2)*, 198–202.

Phelps, E. A., LeDoux, J. E. (2005). Contributions of the amygdala to emotion processing: from animal models to human behavior. *Neuron, 48(2)*, 175–187.

Phelps, E. A., O'Connor, K. J., Gatenby, J. C., Gore, J. C., Grillon, C., Davis, M. (2001). Activation of the left amygdala to a cognitive representation of fear. *Nature Neuroscience, 4(4)*, 437–441.

Piaget, J. (1954). *Intelligenz und Affektivität in der Entwicklung des Kindes.* Frankfurt am Main: Suhrkamp 1995.

Pickel, A. (2005). The *habitus* process: A biopsychosocial conception. *Journal for the Theory of Social Behaviour, 35(4)*, 437–461.

Popitz, H. (1980). *Die normative Konstruktion von Gesellschaft.* Tübingen: Mohr.

Prinz, J. (2007). *The emotional construction of morals.* Oxford: Oxford University Press.

de Quervain, D. J.-F., Fischbacher, U., Treyer, V., Schellhammer, M., Schnyder, U., Buck, A., Fehr, E. (2004). The neural basis of altruistic punishment. *Science, 305*, 1254–1258.

Rackow, K., Schupp, J., von Scheve, C. (2012). Angst und Ärger. Zur Relevanz emotionaler Dimensionen sozialer Ungleichheit. *Zeitschrift für Soziologie, 41(5)*, 392–409.

Radden, J. (2000). *The nature of melancholy.* New York: Oxford University Press.

Reay, D. (2000). A useful extension of Bourdieu's conceptual framework? Emotional capital as a way of understanding mothers' involvement in their children's education. *Sociological Review, 48(4)*, 568–585.

Reckwitz, A. (2002). Toward a theory of social practices. A development in culturalist theorizing. *European Journal of Social Theory, 5(2)*, 245–265.

Reddy, W. M. (2001). *The navigation of feeling.* New York: Cambridge University Press.

Reich, W. (2010). Three problems of intersubjectivity—and one solution. *Sociological Theory, 28(1)*, 40–63.

Reichertz, J., Zaboura, N. (2006) (Eds.), *Akteur Gehirn—oder das vermeintliche Ende des handelnden Subjekts.* Wiesbaden: VS Verlag.

Reisenzein, R. (1983). The Schachter theory of emotion: Two decades later. *Psychological Bulletin, 94(2),* 239–264.

Reisenzein, R. (1998). Outlines of a theory of emotions as metarepresentational states of mind. In Fischer, A. H. (Ed.), *Proceedings of the 10th conference of the International Society for Research on Emotions.* Amsterdam: ISRE, 186–191.

Reisenzein, R. (2000). Einschätzungstheoretische Ansätze in der Emotionspsychologie. In Otto, J. H., Euler, H. A., Mandl, H. (Eds.), *Handbuch der Emotionspsychologie.* Weinheim: PsychologieVerlagsUnion, 117–138.

Reisenzein, R. (2001). Appraisal processes conceptualized from a schema theoretic perspective: Contributions to a process analysis of emotions. In Scherer, K. R., Schorr, A., Johnstone, T. (Eds.), *Appraisal processes in emotion.* New York: Oxford University Press, 187–204.

Resnick, L. B. (1991). Shared cognition: Thinking as social practice. In Resnick, L. B., Levine, J. M., Teasley, S. D. (Eds.), *Perspectives on socially shared cognition.* Washington, DC: American Psychological Association, 1–22.

Rilling, J. K., Gutman, D. A., Zeh, T. R., Pagnoni, G., Berns, G. S., Kilts, C. D. (2002). A neural basis for social cooperation. *Neuron, 35(2),* 395–405.

Rimé, B. (2009). Emotion elicits the social sharing of emotion: Theory and empirical review. *Emotion Review, 1,* 60–85.

Rivera, J. de (1992). Emotional climate: Social structure and emotional dynamics. In Strongman, K. T. (Ed.), *International review of studies on emotion.* Vol. 2. Chichester, UK: Wiley, 197–218.

Robinson, M. D. (1998). Running from William James' bear: A review of preattentive mechanisms and their contributions to emotional experience. *Cognition and Emotion, 12(5),* 667–696.

Rolls, E. T. (1990). A theory of emotion, and its application to understanding the neural basis of emotion. *Cognition and Emotion, 4(3),* 161–190.

Rolls, E. T. (1999). *The brain and emotion.* Oxford: Oxford University Press.

Rolls, E. T. (2000). Precis of "The brain and emotion." *Behavioral and Brain Sciences, 23(2),* 177–234.

Rolls, E. T. (2002). Emotion, neural basis of. In Smelser, N. J., Baltes, P. B. (Eds.), *International encyclopedia of the social and behavioral sciences.* Amsterdam: Pergamon Press, 4444–4449.

Rolls, E. T. (2004). The functions of the OFC. *Brain and Cognition, 55(1),* 11–29.

Roseman, I. J. (1991). Appraisal determinants of discrete emotions. *Cognition and Emotion, 5(3),* 161–200.

Roseman, I. J. (2001). A model of appraisal in the emotion system. Integrating theory, research, and applications. In In Scherer, K. R., Schorr, A., Johnstone, T. (Eds.), *Appraisal processes in emotion.* New York: Oxford University Press, 68–91.

Roseman, I. J., Smith, C. A. (2001). Appraisal theory: Overview, assumptions, varieties, controversies. In Scherer, K. R., Schorr, A., Johnstone, T. (Eds.), *Appraisal processes in emotion.* New York: Oxford University Press, 3–19.

Roseman, I. J., Antoniou, A. A., Jose, P. E. (1996). Appraisal determinants of emotions: Constructing a more accurate and comprehensive theory. *Cognition and Emotion, 10(3),* 241–277.

Roth, G. (2003). *Fühlen, Denken, Handeln.* Frankfurt am Main: Suhrkamp.

Roy, J. P. (2004). Socioeconomic status and health: a neurobiological perspective. *Medical Hypotheses, 62(2)*, 222–227.

Russell, J. A. (1994). Is there universal recognition of emotion from facial expression? A review of the cross-cultural studies. *Psychological Bulletin, 115(1)*, 102–141.

Russell, J. A. (1995). Facial expressions of emotions: What lies beyond minimal universality? *Psychological Bulletin, 118(3)*, 379–391.

Russell, J. A. (2003). Core affect and the psychological construction of emotion. *Psychological Review, 110*, 145–172.

Russell, J. A. (2009). Emotion, core affect, and psychological construction. *Cognition and Emotion, 23(7)*, 1259–1283.

Russell, J. A., Fernandez-Dols, J.-M. (1997). What does a facial expression mean? In Russell, J. A., Fernandez-Dols, J.-M. (Eds.), *The psychology of facial expression*. New York: Cambridge University Press, 3–30.

Russell, J. A., Bachorowski, J.-A., Fernandez-Dols, J.-M. (2003). Facial and vocal expressions of emotion. *Annual Review of Psychology, 54*, 329–349.

Sander, D., Grafman, J., Zalla, T. (2003). The human amygdala: An evolved system for relevance detection. *Reviews in the Neurosciences, 14*, 303–316.

Sanfey, A. G., Rilling, J. K., Aronson, J. A., Nystrom, L. E., Cohen, J. D. (2003). The neural basis of economic decision-making in the ultimatum game. *Science, 300*, 1755–1758.

Schachter, S., Singer, J. E. (1962). Cognitive, social, and physiological determinants of emotional state. *Psychological Review, 69(5)*, 379–399.

Schaefer, M., Rotte, M. (2007). Favorite brands as cultural objects modulate reward circuit. *Neuroreport, 18(2)*, 141–145.

Schatzki, T. R. (1996). *Social Practices. A Wittgensteinian approach to human activity and the social*. Cambridge: Cambridge University Press.

Scheff, T. J. (1997). *Emotions, the social bond, and human reality*. Cambridge: Cambridge University Press.

Scheff, T. J. (2000). Shame and the social bond. A sociological theory. *Sociological Theory, 18(1)*, 84–99.

Scheff, T. J., Retzinger, S. M. (2000). Shame as the master emotion of everyday life. *Journal of Mundane Behavior, 1(3)*. Retrieved August 23, 2004, from www.mundanebehavior.org/issues/v1n3/scheff-retzinger.htm.

Scherer, K. R. (1984). On the nature and function of emotion: A component process approach. In Scherer, K. R., Ekman, P. (Eds.), *Approaches to emotion*. Hillsdale, NJ: Erlbaum, 293–318.

Scherer, K. R. (1993a). Neuroscience projections to current debates in emotion psychology. *Cognition and Emotion, 7(1)*, 1–41.

Scherer, K. R. (1993b). Studying the emotion-antecedent appraisal process: An expert system approach. *Cognition and Emotion, 7(3/4)*, 325–355.

Scherer, K. R. (1994). Emotion serves to decouple stimulus and response. In Ekman, P., Davidson, R. J. (Eds.), *The nature of emotion*. New York: Oxford University Press, 127–130.

Scherer, K. R. (1997). The role of culture in emotion-antecedent appraisal. *Journal of Personality and Social Psychology, 73(5)*, 902–922.

Scherer, K. R. (1999). On the sequential nature of appraisal processes: Indirect evidence from a recognition task. *Cognition and Emotion, 13(6)*, 763–793.

Scherer, K. R. (2005). What are emotions? And how can they be measured? *Social Science Information, 44(4)*, 695–729.

von Scheve, C. (2012a). Emotion work and emotion regulation: Two sides of the same coin? *Frontiers in Emotion Science, 3*, 496. doi: 10.3389/fpsyg.2012.00496.

von Scheve, C. (2012b). The social calibration of emotion expression: An affective basis of micro-social order. *Sociological Theory, 30(1)*, 1–14.

von Scheve, C., Ismer, S. (2013). Towards a theory of collective emotions. *Emotion Review, 5(4)*, 7–8.

Schieman, S. (2003). Socioeconomic status and the frequency of anger across the life course. *Sociological Perspectives, 46(2)*, 207–222.

Schmidt-Atzert, L. (1996). *Lehrbuch der Emotionspsychologie*. Stuttgart: Kohlhammer.

Schütz, A. (1953). Common-sense and scientific interpretation of human action. *Philosophy and Phenomenological Research, 14(1)*, 1–38.

Schütz, A., Luckmann, T. (1973). *The structures of the life-world*. Evanston, IL: Northwestern University Press.

Schwarz, N., Bless, H. (1991). Happy and mindless, but sad and smart? The impact of affective states on analytic reasoning. In Forgas, J. P. (Ed.), *Emotion and social judgments*. Oxford: Pergamon Press, 55–72.

Schwarz, N., Clore, G. L. (1988). How do I feel about it? The informative function of affective states. In Fiedler, K., Forgas, J. P. (Eds.), *Affect, cognition, and social behavior*. Toronto: Hogrefe, 44–62.

Searle, J. R. (1990). Collective intentions and actions. In Cohen, P., Morgan, J., Pollack, M. (Eds.), *Intentions in communication*. Cambridge, MA: MIT Press, 401–415.

Searle, J. R. (1995). *The construction of social reality*. New York: Free Press.

Shore, B. (1996). *Culture in mind*. New York: Oxford University Press.

Simmel, G. (1901). Zur Psychologie der Scham. In Dahme, H.-J., Rammstedt, O. (Eds.), *Schriften zur Soziologie*. Frankfurt am Main: Suhrkamp, 151–158.

Sloman, S. A. (1996). The empirical case for two systems of reasoning. *Psychological Bulletin, 119(1)*, 3–22.

Smelser, N. J. (1992). The rational choice perspective. A theoretical assessment. *Rationality and Society, 4*, 381–410.

Smith, C. A., Ellsworth, P. C. (1985). Patterns of cognitive appraisal in emotion. *Journal of Personality and Social Psychology, 48(4)*, 813–838.

Smith, C. A., Kirby, L. D. (2000). Consequences require antecedents: Toward a process model of emotion elicitation. In Forgas, J. P. (Ed.), *Feeling and thinking. The role of affect in social cognition*. New York: Cambridge University Press, 83–106.

Smith, C. A., Kirby, L. D. (2001). Toward delivering on the promise of appraisal theory. In Scherer, K. R., Schorr, A., Johnstone, T. (Eds.), *Appraisal processes in emotion*. New York: Oxford University Press, 121–138.

Smith, C. A., Lazarus, R. S. (1993). Appraisal components, core relational schemes, and the emotions. *Cognition and Emotion, 7(3/4)*, 233–269.

Smith, C. A., David, B., Kirby, L. (2006). Emotion-eliciting appraisals of social situations. In Forgas, J. P. (Ed.), *Affect in social thinking and behavior*. New York: Psychology Press, 85–101.

Smith, E. R., Queller, S. (2004). Mental representations. In Brewer, M. B., Hewstone, M. (Eds.), *Social cognition*. Malden, MA: Blackwell, 5–27.

Solomon, R. C. (1976). *The passions*. New York: Doubleday-Anchor.

Solomon, R. C. (2004). On the passivity of the passions. In Manstead, A. S., Frijda, N. H., Fischer, A. (Eds.), *Feelings and emotions*. New York: Oxford University Press, 11–29.

de Sousa, R. (1990). *The rationality of emotion*. Cambridge, MA: MIT Press.

de Sousa, R. (2010). Emotion. *Stanford Encyclopedia of Philosophy*. Retrieved April 20, 2011, from http://plato.stanford.edu/entries/emotion.

Sperber, D., Hirschfeld, L. A. (2004). The cognitive foundations of cultural stability and diversity. *Trends in Cognitive Sciences, 8(1)*, 40–46.

Squire, L. R. (2004). Memory systems of the brain: A brief history and current perspective. *Neurobiology of Learning and Memory, 82(3)*, 171–177.

Stearns, P. N. (1994). *American cool. Constructing a twentieth-century emotional style*. New York: New York University Press.

Stein, N. L., Oatley, K. (1992). Basic emotions: Theory and measurement. *Cognition and Emotion, 6(3/4)*, 161–168.

Stephen, A. T., Pham, M. T. (2008). On feelings as a heuristic for making offers in ultimatum negotiations. *Psychological Science, 19(10)*, 1051–1058.

Stolte, J. F., Fine, G. A., Cook, K. S. (2001). Sociological miniaturism: Seeing the big through the small in social psychology. *Annual Review of Sociology, 27*, 387–413.

Storbeck, J., Clore, G. L. (2007). On the interdependence of cognition and emotion. *Cognition and Emotion, 21(6)*, 1212–1237.

Summers-Effler, E. (2002). The micro potential for social change: Emotion, consciousness, and social movement formation. *Sociological Theory, 20(1)*, 41–60.

Tamir, M. (2009). What do people want to feel and why? Pleasure and utility in emotion regulation. *Current Directions in Psychological Science, 18(2)*, 101–105.

Tamir, M., Mauss, I. B. (2011). Social-cognitive factors in emotion regulation: Implications for well-being. In Nyklicek, I., Vingerhoets, A., Zeelenberg, M., Denollet, J. (Eds.), *Emotion regulation and well-being*. New York: Springer, 31–47.

Teasdale, J. D., Howard, R. J., Cox, S. G., Ha, Y., Brammer, M. J., Williams, S. C., Checkley, S. A. (1999). Functional MRI study of the cognitive generation of affect. *American Journal of Psychiatry, 156(2)*, 209–215.

TenHouten, W. D. (2007). *A general theory of emotions and social life*. New York: Routledge.

Thoits, P. A. (1990). Emotional deviance: Research agendas. In Kemper, T. D. (Ed.), *Research agendas in the sociology of emotions*. Albany, NY: State University of New York Press, 180–203.

Thoits, P. A. (2004). Emotion norms, emotion work, and social order. In Manstead, A. S., Frijda, N. H., Fischer, A. (Eds.), *Feelings and emotions*. New York: Oxford University Press, 359–378.

Thoits, P. A. (2007). Extending Scherer's conception of emotion. *Social Science Information, 46*, 429–433.

Tomkins, S. S. (1962). *Affect, imagery, consciousness*. Vol. 1. New York: Springer.

Tracy, J. L., Matsumoto, D. (2008). The spontaneous expression of pride and shame: Evidence for biologically innate nonverbal displays. *Proceedings of the National Academy of Sciences, 105*, 11655–11660.

Tsai, J. L. (2007). Ideal affect. Cultural causes and behavioral consequences. *Perspectives on Psychological Science, 2(3)*, 242–259.

Tsai, J. L., Knutson, B., Fung, H. H. (2006). Cultural variation in affect valuation. *Journal of Personality and Social Psychology, 90*, 288–307.

Tuomela, R. (1995). *The importance of us*. Stanford, CA: Stanford University Press.

Turner, J. H. (1988). A behavioral theory of social structure. *Journal for the Theory of Social Behaviour, 18(4)*, 354–372.

Turner, J. H. (1999a). The neurology of emotion. Implications for sociological theories of interpersonal behavior. In Franks, D. D., Smith, T. S. (Eds.), *Mind, brain, and society: Toward a neurosociology of emotion*. Greenwich, CT: JAI Press, 81–108.

Turner, J. H. (1999b). Toward a general sociological theory of emotions. *Journal for the Theory of Social Behaviour, 29(2)*, 133–161.

Turner, J. H. (2000). *On the origins of human emotions*. Stanford, CA: Stanford University Press.

Turner, J. H. (2002). *Face to face*. Stanford, CA: Stanford University Press.

Turner, J. H. (2007). *Human emotions. A sociological theory*. London: Routledge.

Turner, J. H. (2011). *The problem of emotions in societies*. New York: Routledge.

Turner, J. H., Collins, R. (1989). Toward a microtheory of structuring. In Turner, J. H. (Ed.), *Theory building in sociology. Assessing theoretical cumulation*. Newbury Park, CA: Sage, 118–130.

Turner, R. H. (1978). The role and the person. *American Journal of Sociology, 84(1)*, 1–23.

Turner, S. P. (1994). *The social theory of practices*. Cambridge: Polity.

Turner, S. P. (2002). *Brains/practices/relativism: Social theory after cognitive science*. Chicago, IL: University of Chicago Press.

Turner, S. P. (2007). Social theory as a cognitive neuroscience. *European Journal of Social Theory, 10(3)*, 357–374.

Vastfjall, D., Garling, T., Kleiner, M. (2001). Does it make you happy feeling this way? A core affect account of preferences for current mood. *Journal of Happiness Studies, 2*, 337–354.

Vygotsky, S. (1978). *Mind in society: Development of higher psychological processes*. Cambridge, MA: Harvard University Press.

Weber, M. (1968). *Economy and Society*. Berkeley, CA: University of California Press.

Weber, M. (1991). The nature of social action. In Runciman, W. G. (Ed.), *Weber: Selections in translation*. Cambridge: Cambridge University Press.

Weise, P. (1989). Homo oeconomicus und homo sociologicus. Die Schreckensmänner der Sozialwissenschaften. *Zeitschrift für Soziologie, 18(2)*, 148–161.

Weiss, H. M., Brief, A. P. (2001). Affect at work: An historical perspective. In Payne, R. L., Cooper, C. L. (Eds.), *Emotions at work*. Chichester, UK: Wiley, 133–172.

Welzer, H., Markowitsch, H. J. (2001). Umrisse einer interdisziplinären Gedächtnisforschung. *Psychologische Rundschau, 52(4)*, 205–214.

Wertsch, J. V. (1991). A sociocultural approach to socially shared cognition. In Resnick, L. B., Levine, J. M., Teasley, S. D. (Eds.), *Perspectives on socially shared cognition*. Washington, DC: American Psychological Association, 85–100.

Wilkinson, R. G. (1999). Health, hierarchy, and social anxiety. *Annals of the New York Academy of Sciences, 896*, 48–63.

Wilkinson, R. G., Pickett, K. (2009). *The spirit level: Why more equal societies almost always do better*. London: Allen Lane.

Wilson, M. (2002). Six views of embodied cognition. *Psychonomic Bulletin & Review, 9(4)*, 625–636.

Winkielman, P., Knutson, B., Paulus, M., Trujillo, J. L. (2007). Affective influence on judgments and decisions: Moving towards core mechanisms. *Review of General Psychology, 11(2)*, 179–192.

Wood, J. N., Grafman, J. (2003). Human prefrontal cortex: Processing and representational perspectives. *Nature Reviews Neuroscience, 4*, 139–147.

Wrong, D. (1961). The oversocialized conception of man in modern sociology. *American Sociological Review, 26,* 184–193.

Yeung, C. W., Wyer, R. S. (2004). Affect, appraisal, and consumer judgment. *Journal of Consumer Research, 31(2),* 412–424.

Young, H. P. (2008). Social norms. In Durlauf, S. N., Blume, L. E. (Eds.), *The new Palgrave dictionary of economics.* 2nd ed. Palgrave Macmillan. Available at www.dictionaryofeconomics.com.

Zaalberg, R., Manstead, A. S. R., Fischer, A. H. (2004). Relations between emotions, display rules, social motives, and facial behavior. *Cognition and Emotion, 18(2),* 183–207.

Zajonc, R. (1980). Feeling and thinking: Preferences need no inferences. *American Psychologist, 35(2),* 151–175.

Zerubavel, E. (1997). *Social mindscapes.* Cambridge, MA: Harvard University Press.

Zink, C. F., Tong, Y., Chen, Q., Bassett, D. S., Stein, J. L., Meyer-Lindenberg, A. (2008). Know your place: neural processing of stable and unstable social hierarchy in humans. *Neuron, 58,* 273–283.

Index

Page references to figures and tables are in *italics* and those for notes are followed by the letter 'n'.